Construction Extension to The PMBOK® Guide Third Edition

ISBN: 978-1-930699-52-6

Published by: Project Management Institute, Inc.
14 Campus Boulevard
Newtown Square, Pennsylvania 19073-3299 USA.
Phone: +610-356-4600
Fax: +610-356-4647
E-mail: customercare@pmi.org
Internet: www.pmi.org

PMI Publications welcomes corrections and comments on its books. Please feel free to send comments on typographical, formatting, or other errors. Simply make a copy of the relevant page of the book, mark the error, and send it to: Book Editor, PMI Publications, Four Campus Boulevard, Newtown Square, PA 19073-3299 USA.

To inquire about discounts for resale or educational purposes, please contact the PMI Book Service Center.

PMI Book Service Center
P.O. Box 932683, Atlanta, GA 31193-2683 USA
Phone: 1-866-276-4764 (within the U.S. or Canada) or +1-770-280-4129 (globally)
Fax: +1-770-280-4113
E-mail: info@bookorders.pmi.org

The paper used in this book complies with the Permanent Paper Standard issued by the National Information Standards Organization (Z39.48—1984).

10 9 8 7 6 5

Notice

The Project Management Institute, Inc. (PMI) standards and guideline publications, of which the document contained herein is one, are developed through a voluntary consensus standards development process. This process brings together volunteers and/or seeks out the views of persons who have an interest in the topic covered by this publication. While PMI administers the process and establishes rules to promote fairness in the development of consensus, it does not write the document and it does not independently test, evaluate, or verify the accuracy or completeness of any information or the soundness of any judgments contained in its standards and guideline publications.

PMI disclaims liability for any personal injury, property or other damages of any nature whatsoever, whether special, indirect, consequential or compensatory, directly or indirectly resulting from the publication, use of application, or reliance on this document. PMI disclaims and makes no guaranty or warranty, expressed or implied, as to the accuracy or completeness of any information published herein, and disclaims and makes no warranty that the information in this document will fulfill any of your particular purposes or needs. PMI does not undertake to guarantee the performance of any individual manufacturer or seller's products or services by virtue of this standard or guide.

In publishing and making this document available, PMI is not undertaking to render professional or other services for or on behalf of any person or entity, nor is PMI undertaking to perform any duty owed by any person or entity to someone else. Anyone using this document should rely on his or her own independent judgment or, as appropriate, seek the advice of a competent professional in determining the exercise of reasonable care in any given circumstances. Information and other standards on the topic covered by this publication may be available from other sources, which the user may wish to consult for additional views or information not covered by this publication.

PMI has no power, nor does it undertake to police or enforce compliance with the contents of this document. PMI does not certify, test, or inspect products, designs, or installations for safety or health purposes. Any certification or other statement of compliance with any health or safety-related information in this document shall not be attributable to PMI and is solely the responsibility of the certifier or maker of the statement.

Contents

Preface

In 2002, PMI began publishing industry-specific application area extensions to *A Guide to the Project Management Body of Knowledge* (*PMBOK® Guide*). The *Construction Extension to the PMBOK® Guide 2000 Edition* (provisional version, published in 2003) was the second such application area extension for a specific industry area.

With the publication of the *PMBOK® Guide*—Third Edition in 2004, it became essential to update these industry extensions to maintain full consistency with the changes in the new standard. This new extension, *Construction Extension to the PMBOK® Guide Third Edition,* supersedes the *Construction Extension to the PMBOK® Guide 2000 Edition* and is aligned with, and a supplement to, the *PMBOK® Guide*—Third Edition. The alignment enables easier reference to the corresponding section in each document. The process names and designations were updated to match the changes introduced in the current edition of *PMBOK® Guide* in order to enable consistency and clarity.

The *PMBOK® Guide*—Third Edition describes knowledge and practices "generally recognized as good practices on most projects most of the time." The *Construction Extension to the PMBOK® Guide Third Edition* describes knowledge and practices that are "generally accepted as good practices' for "most construction projects most of the time." As an extension, this standard builds upon the *PMBOK® Guide*—Third Edition, by describing additional knowledge and practices and by modifying some of them. It is intended to be completely consistent with that standard. The *Construction Extension to the PMBOK® Guide Third Edition* also includes Knowledge Areas (Chapters 13–16) that are specific to the construction industry, which are not in the *PMBOK® Guide*—Third Edition because they do not apply to most projects most of the time. These construction industry unique Knowledge Areas are included at the end of Section III.

Additional figures were added to Chapter 1 of the *Construction Extension to the PMBOK® Guide Third Edition* which do not appear in the *PMBOK® Guide*—Third Edition.

Section I

The Project Management Framework

Chapter 1

Introduction

A Guide to the Project Management Body of Knowledge (*PMBOK® Guide*)—Third Edition describes the Project Management Body of Knowledge as "the sum of knowledge within the profession of project management" that resides with practitioners and academics who apply and advance it. While the *PMBOK® Guide*—Third Edition provides a generic foundation for managing projects, this extension addresses the specific practices found in construction projects. This document supersedes the previous edition of the *Construction Extension*, entitled *Construction Extension to the PMBOK® Guide 2000 Edition*.

This extension to the *PMBOK® Guide*—Third Edition describes the generally accepted principles for construction projects that are not common to all project types. The general organization scheme of Knowledge Areas and processes found in the *PMBOK® Guide*—Third Edition is also used in this extension for consistency and ease of use. Project management professionals should plan to use both documents concurrently in the execution of their responsibilities. To that end, this chapter is organized into the following sections which reflect a similar structure to the *PMBOK® Guide*—Third Edition:

- **1.1 Purpose of the *Construction Extension***
- **1.2 What Is a Project?**
- **1.3 Project Management in the Construction Context**
- **1.4 Structure of the *Construction Extension***
- **1.5 Areas of Expertise**
- **1.6 Project Management Context**
- **1.7 Explanation of *Construction Extension* Processes: Inputs, Tools & Techniques, and Outputs**

1.1 Purpose of the Construction Extension

The primary purpose of *A Guide to the Project Management Body of Knowledge* is "to identify that subset of the body of knowledge that is generally recognized as good practice."

Appendix D of the *PMBOK® Guide*—Third Edition describes application area extensions as follows:

"Application area extensions are necessary when there are generally accepted knowledge and practices for a category of projects in one application area that are not generally accepted across the full range of project types in most application areas. Application area extensions reflect:

- Unique or unusual aspects of the project environment for which the project management team must be aware in order to manage the project efficiently and effectively
- Common knowledge and practices that, if followed, will improve the efficiency and effectiveness of the project (e.g., standard work breakdown structures)."

This standard is an application area extension for construction projects. The key characteristics of these projects are listed in Section 1.2. The purpose of this extension is to improve the efficiency and effectiveness of the management of construction projects and to include material specifically applicable to construction that is not presently covered in the *PMBOK® Guide*—Third Edition.

Much of the *PMBOK® Guide*—Third Edition is directly applicable to construction projects. In fact, the practices and project management of construction projects were one of the foundations of the original 1987 document, *The Project Management Body of Knowledge*. Since that time, a growing awareness of the value of project management to all kinds of projects and industries has led to a broadening of concepts and an inclusiveness that, because of its more universal nature, does not, in some aspects, fully cover present-day project management practices found in the worldwide construction industry. For this reason, while the changes may not be substantial, they are different enough from other industries and applications to warrant an extension.

1.1.1 Audience for the *Construction Extension to the PMBOK® Guide Third Edition*

The audience for the *Construction Extension to the PMBOK® Guide Third Edition* includes but is not limited to:

- Taxpayers
- Regulatory agencies
- Local government groups
- Environmental groups
- Community groups
- Risk management specialists
- Civil engineers
- Architectural and engineers
- Geotechnical experts
- Financial specialists
- Construction project managers
- Contractors
- Construction industry tradesmen and professionals
- Other stakeholders in the construction process, from land acquisition through design, construction, and occupancy

1.2 What Is a Project?

See Section 1.2 of the *PMBOK® Guide*—Third Edition.

1.2.1 Project Characteristics

See Section 1.2.1 of the *PMBOK® Guide*—Third Edition.

1.2.2 Projects vs. Operation Work

See Section 1.2.2 of the *PMBOK® Guide*—Third Edition.

1.2.3 Projects and Strategic Planning

See Section 1.2.3 of the *PMBOK® Guide*—Third Edition.

1.2.4 Why are Construction Projects Unique?

Construction projects inherently contain a high degree of risk in their projections of cost and time as each is unique. Buildings may be prototypical, but when constructed on different sites, each project presents its own challenges to accurate cost, time projections, and control. Construction projects in the industrial sector will quite often require intricate interface with technology licensors that demand construction techniques be varied to suit the nuances of their technology transfer. This can contribute to unique subcontracting arrangements, extensions to schedule, and increases in capital cost.

Construction projects must address the geography and conditions of the project site and the relation of the project to the environment. Construction projects often result in one-off products rather than mass-produced products. While there is generally no opportunity to produce a prototype, a construction project may sometimes be phased to provide an opportunity to refine the project design in the initial phase.

Construction projects produce deliverables, such as: a facility that will make or house the means to make a product or provide service facilities such as dams, highways, parks, institutions, entire developments (for example, high rises and educational, military housing, or airports), or infrastructures that deliver water, electricity, telecommunications, or wastewater disposal. Other examples are schools, medical centers, and hospitals.

Construction projects often are required to have, by regulations, a team of hired specialists and construction disciplines involved on the project.

In today's world, construction projects involve many stakeholders with varying project expectations such as public taxpayers, regulatory agencies, governments, and environmental or community groups, which many other types of projects do not include.

Construction projects often require large amounts of materials and physical tools to move or modify those materials.

1.3 Project Management in the Construction Context

See Section 1.3 of the *PMBOK® Guide*—Third Edition.

The nine Knowledge Areas in the *PMBOK® Guide*—Third Edition are all applicable to construction projects. However, in this extension, they have been modified to address the unique attributes that are specific to the construction industry and to emphasize those activities that are important in construction. There are additional important Knowledge Areas that apply specifically to construction projects as follows:

- Safety Management
- Environmental Management
- Financial Management
- Claim Management

While some aspects of these Knowledge Areas may be found in portions of the nine basic Knowledge Areas, their importance and universality in construction necessitate their addition as Knowledge Areas in this extension.

1.4 Structure of the Construction Extension

See also Section 1.4 of the *PMBOK® Guide*—Third Edition.

The *Construction Extension to the PMBOK® Guide*—Third Edition is organized into the following three sections:

- The Project Management Framework
- The Standard for Project Management of a Project
- The Project Management Knowledge Areas

1.4.1 Section I—The Project Management Framework

See also Section 1.4.1 of the *PMBOK® Guide*—Third Edition.

Specific sections of Chapter 1 (Introduction) and Chapter 2 (Project Life Cycle and Organization) of this extension describe features that are specific to construction projects; otherwise the reader is referred to the appropriate section(s) of the *PMBOK® Guide*—Third Edition.

1.4.2 Section II—The Standard for Project Management of a Project

See also Section 1.4.2 of the *PMBOK® Guide*—Third Edition. Chapter 3 (Project Management Processes for a Project) of this extension describes project management processes that are used by a project team to manage a construction project.

1.4.3 Section III—The Project Management Knowledge Areas

See also Section 1.4.3 of the *PMBOK® Guide*—Third Edition.

Chapters 4 through 12 correspond to the nine Knowledge Areas described in the *PMBOK® Guide*—Third Edition with specific additions or modifications that are related to construction projects (see overview in Figure 1-1). The *Construction Extension* includes four added Knowledge Areas specifically applicable to construction projects (see Figure 1-2):

Chapter 13—Project Safety Management describes the processes required to ensure that the construction project is executed with appropriate care to prevent accidents that could cause personal injury or property damage. This Knowledge Area includes

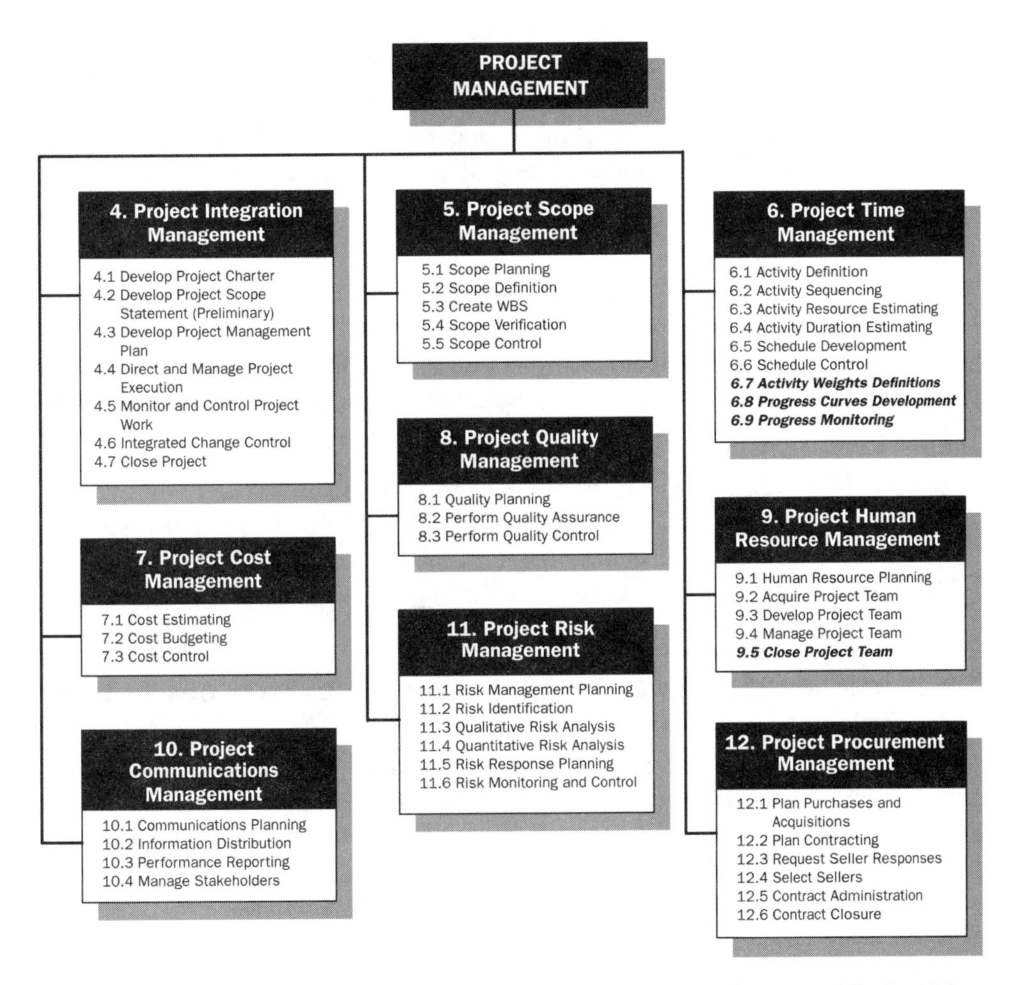

Figure 1-1. Overview of Project Management Knowledge Areas and Project Management Processes

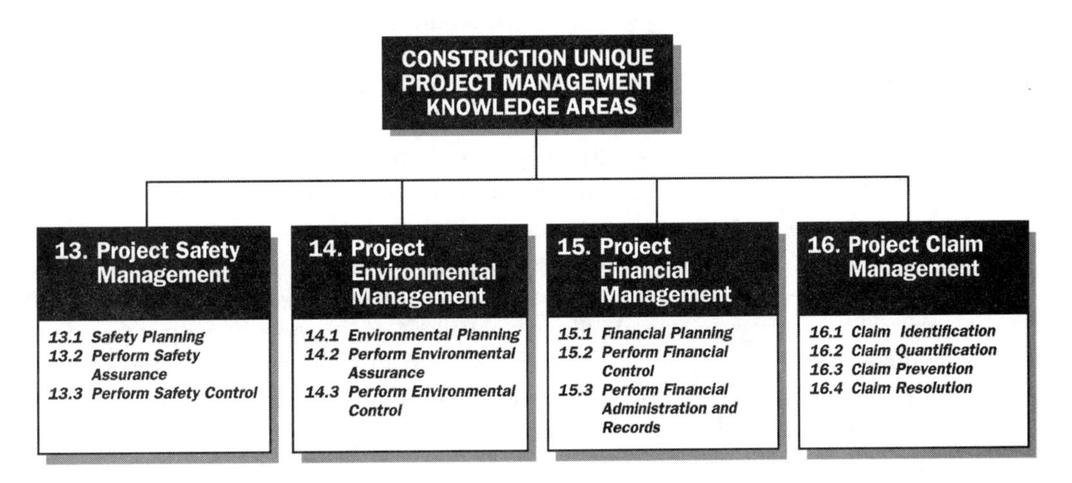

Figure 1-2. Overview of Project Management Knowledge Areas and Project Management Processes Specific to Construction Projects

the safety and health of the general public who may be exposed to the project. Safety Management also includes human resource programs such as drug and alcohol treatment programs, which contribute to reducing accidents on construction jobsites.

Chapter 14—Project Environmental Management describes the processes required to ensure that the impact of the project on the surrounding environment remains within the limits stated in the project's legal permits. Environmental management also addresses the health and welfare of the workers and the occupants at a facility during and after construction.

Chapter 15—Project Financial Management describes the processes needed to acquire and manage the financial resources for the project. Compared to cost management, this knowledge area is more concerned with revenue sources and analyzing net cash flows for the project than with day-to-day cost controls. In this chapter, the discussion is limited to financing the cost of construction for the project itself.

Chapter 16—Project Claim Management describes the processes required to prevent construction claims and to expedite the handling of claims when they do occur. This chapter presents an outline view of claim management to encourage a careful approach to contract preparation.

1.4.4 Other Management Systems Standards Applicable to Construction Projects

There are also other management systems standards and guidelines that apply to construction projects. Examples include those developed by the International Organization for Standardization (ISO); the most recognized of these are the ISO 9000 series of standards, the most prevalent of which are:

- ISO 9001 on Quality Management Systems—Requirements
- ISO 9004 on Quality Management Systems—Guidelines for Performance Improvements

Other ISO standards address environmental management, and more recently occupational health and safety management. These ISO standards are those from the ISO 10000, 11000, and 17000 series. These standards break down and further emphasize the principles of the original 9000 series. ISO has developed a vast array of such standards, including those for systems on risk management and communication management.

This extension, however, does not attempt to address all features of the entire range of ISO standards. Rather, it focuses on those that are most commonly associated with construction projects; for example, quality, safety, and environmental management (see Figure 1-3). It should be noted, though, that the approaches to these three areas in this extension (see Chapters 8, 13 and 14, respectively) are intended to be compatible with ISO standards and guidelines, as well as with proprietary approaches to quality management such as those recommended by Deming, Juran, Crosby, and others, and with non-proprietary approaches such as total quality management (TQM), Six Sigma, failure mode and effect analysis, design reviews, voice of the customer, cost of quality (COQ), and continuous improvement, inclusive of quality function deployment (QFD).

Many stakeholders, including those who undertake construction projects, frequently specify ISO standards in their project conditions of contract because the systems developed to comply with these standards can be independently verified and provide stakeholders with additional assurance that the project deliverables are compliant with project requirements. One of the primary values of the standards is that they can be used in contractual situations. It is common for stakeholders to require the management systems of performing organizations to be ISO-compliant. The ISO management system standards and the requirements of this extension are generally com-

Figure 1-3. ISO 9000 Series Management Standards

patible, although there are differences, especially regarding strategic planning, even beyond the requirements defined in Chapters 8, 13, and 14.

.1 **Prevention Over Inspection.** See *PMBOK® Guide*—Third Edition, Chapter 8. Although verification by inspection and testing are synonymous as they pertain to construction projects, it is also important to determine the most appropriate methods of verification and verification criteria. Critical issues include but are not limited to:

- Evaluating performance characteristics of the end product of the project (or component part) as a whole vs. inspecting or testing individual characteristics.
- Ensuring that the correct and objective data is captured, rather than arbitrary or subjective data.

.2 **Management Responsibility.** See *PMBOK® Guide*—Third Edition, Chapter 8. Management responsibility includes the timely provision of adequate resources, in terms of finance, human resources, and equipment, etc.

.3 **Continuous Improvement.** See *PMBOK® Guide*—Third Edition, Chapter 8. Continuous improvement is integral to project management, and needs to be integrated into the project management system as a whole.

.4 **Risk Management.** Quality, safety, and environmental management are frequently considered a subset of risk management. As such, these must address more than the technical requirements of a project. It is frequently overlooked that the contract itself is a quality, safety, and environmental management system standard. A specialized management system to address the unique characteristics of a specific construction project is frequently required. For example, the requirements for surface rail or highway infrastructure projects are fundamentally different. Such specialization would frequently require the adoption of both the requirements of this extension and ISO standards.

1.5 Areas of Expertise

See Section 1.5 of the *PMBOK® Guide*—Third Edition. Understanding and applying the knowledge, skills, tools, and techniques which are generally recognized as good practice, is not sufficient alone for effective project management.

In construction projects, as with all projects, expertise is woven throughout the entire process from the start of the project to the finish. For example, the following areas of expertise are interwoven throughout a construction project: design development, procurement of a contractor, individual disciplines on the job site, construction of the project, and the final operation and maintenance of the project. Both the terminology and the means and methods of execution of a project are specific to the type of project, for example, the terminology and execution of tilt-up concrete construction is different from that of residential construction terminology and execution.

1.5.1 Project Management Body of Knowledge

See Section 1.5.1 of the *PMBOK® Guide*—Third Edition. The Knowledge Areas of project management described in the *PMBOK® Guide*—Third Edition can be expanded to include the four additional Knowledge Areas (Chapters 13 to 16) provided in this *Construction Extension.*

1.5.2 Application Area Knowledge, Standards, and Regulations

See Section 1.5.2 of the *PMBOK® Guide*—Third Edition.

1.5.3 Understanding the Project Environment

See Section 1.5.3 of the *PMBOK® Guide*—Third Edition. All types of socio-economic influences are found in construction projects. It is important to maintain a sensitivity and responsiveness to environmental and community concerns as well as mandated regulations. Contracting entities that have government-sponsored business development programs can become an integral part of the construction industry whenever public funds are involved and will influence how a project is implemented.

For companies operating outside of their home country, understanding international and local rules, customs, and cultural differences is particularly important. Virtually all projects are planned and implemented in a social, economic, and environmental context, and have intended and unintended positive and/or negative impacts. Organizations are increasingly accountable for impacts resulting from a project as well as for the effects of a project on people, the economy, and the environment long after it has been completed.

1.5.4 General Management Knowledge and Skills

See Section 1.5.4 of the *PMBOK® Guide*—Third Edition. All of the general management activities mentioned in this section are an integral part of the management of construction projects. General management skills provide much of the foundation for learning project management skills and are essential skills for a project manager. Some general management skills may be relevant only on certain projects or in certain application areas. As an example, team member safety is critical on virtually all construction projects and of little concern on most software development projects.

1.5.5 Interpersonal Skills

See Section 1.5.5 of the *PMBOK® Guide*—Third Edition. In particular, the following interpersonal skills are important for construction projects:

- **Leading.** On a project, the project manager is generally expected to be the leader. Leadership is not, however, limited to the project manager, and may be demonstrated by different individuals at different times throughout the duration of the project. In a construction project, the construction manager is frequently the on-site manager and the project manager interfaces with executive management on schedule, cost, and scope only but not on on-site construction issues.
- **Communicating.** Project communications management is the application of a communication protocol for the specific needs of a project, for example, deciding how, when, in what form, and to whom project performance should be reported. In construction, an example would be the request for information (RFI), which is a request for information on drawings or specifications, or the request for proposal (RFP) from a contractor, which can occur before or during construction.
- **Negotiating.** Negotiating involves conferring with others to come to terms with them or reach an agreement. In construction, negotiating occurs around many issues. Modifications to the contract and negotiating the proposal costs are a few examples of negotiating.
- **Problem Solving.** Problem solving involves a combination of problem definition and decision making. Problem definition requires the distinction between causes and symptoms. Corrective action and preventive action are all part of problem solving. Refer to the glossary for accurate definitions of each and how they apply to problem solving.

All of the skills described are characteristic of the requirements of managing the construction project particularly those involved in negotiating and problem solving.

1.6 Project Management Context

Project management exists in a broader context that includes a program management and project management office. See Section 1.6 of The *PMBOK® Guide*—Third Edition.

1.6.1 Programs and Program Management

See Section 1.6.1 of The *PMBOK® Guide*—Third Edition.

1.6.2 Portfolios and Portfolio Management

See Section 1.6.2 of The *PMBOK® Guide*—Third Edition.

1.6.3 Subprojects

See Section 1.6.3 of The *PMBOK® Guide*—Third Edition.

1.6.4 Project Management Office

See Section 1.6.4 of The *PMBOK® Guide*—Third Edition.

1.7 Explanation of *Construction Extension* Processes: Inputs, Tools and Techniques, and Outputs

This document follows closely the structure and organization of the *PMBOK® Guide*—Third Edition. This enables easier cross-referencing between equivalent sections of this extension and the *PMBOK® Guide.*

The *PMBOK® Guide*—Third Edition describes the inputs, tools and techniques, and outputs of each project management process. For each process, it includes a table that lists three types of elements. This document includes similar tables. In each table, the elements have the following format (see also Figure 1-4):

- Elements that remain unchanged from the *PMBOK® Guide*—Third Edition are shown in plain text
- New items are shown in ***bold italics***
- Changed elements are shown in *italics*

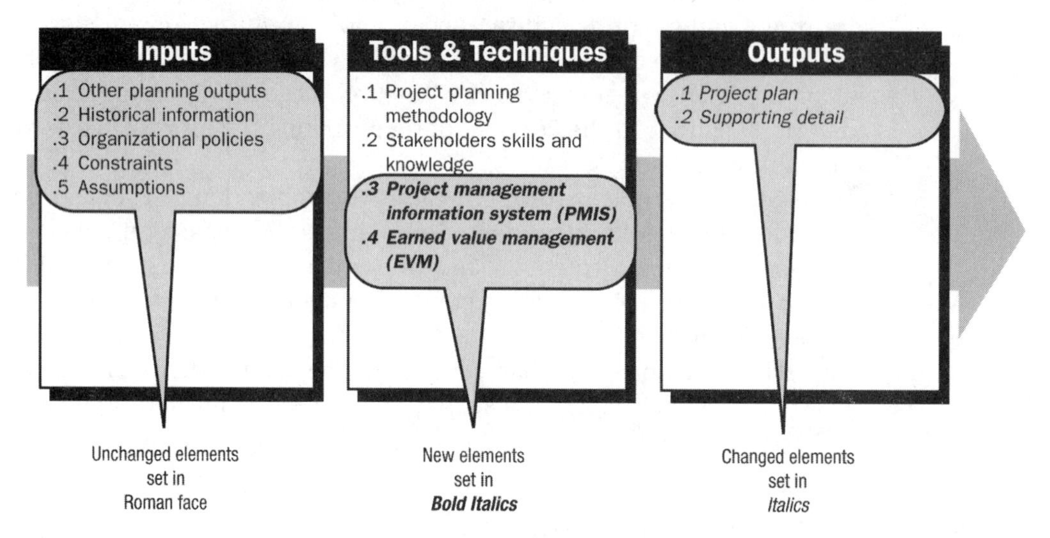

Figure 1-4. Identification of Revised Inputs, Tools & Techniques, and Outputs

Chapter 2

Project Life Cycle and Organization

The *PMBOK® Guide*—Third Edition advises, "Projects and project management are carried out in an environment broader than that of the project itself." Section 1.2.1 of this Construction Extension describes some of the unique features of this environment for construction projects. Most of the contents of this chapter are closely applicable to construction projects.

2.1 The Project Life Cycle

See Section 2.1 of the *PMBOK® Guide*—Third Edition. This concept is particularly important for a construction project where each phase has defined decision points, deliverables, or completion milestones which, when observed, provide a smooth flow and improved control over the life of the project.

2.1.1 Characteristics of the Project Life Cycle

See Section 2.1.1 of the *PMBOK® Guide*—Third Edition. As mentioned in the *PMBOK® Guide,* "The phases of a project life cycle are not the same as the Project Management Process Groups described in detail in Chapter 3."

Figure 2-1 of the *PMBOK® Guide*—Third Edition shows a typical start of a construction project. However, a project proposal could start with design development (DD) or a contractor request for proposal/quotation (RFP/RFQ), depending on the time of initiation of the proposed project.

In the context of project and construction management, the portions of the contracted services that extend beyond the completion of the facility itself (i.e., maintenance or operation) are not considered to be a part of the management of a construction project. While such services may be a part of contracted services and need to be considered, planned, and executed carefully, they take place in the field of "operations" and are not included in the Body of Knowledge related to the management of construction projects.

The most common type of construction project is one that is performed outside of the initiating organization by a team of discipline specialists. However, construction projects which are performed within a single organization do exist and are self-performing entities, such as facility management.

The construction life cycle typically involves three main players: the owner, the designer and the constructor (contractor). Each plays a major role in a construction project, although the responsibilities may vary widely depending on the type of project plan and contract form.

Usually, when the owner determines the need for a new facility or improvement to a facility, a study (often called a feasibility study) is performed to more clearly define the viability and form of the project that will produce the best or most profitable result. The study usually involves a review of alternatives that may satisfy the need (value management) and the desired form of financing (financial management). The study may be conducted in-house if the capability exists. More typically, the services of an architect/engineer(s) or an architect/engineer/contractor are commissioned to do most of the work. While this pre-project activity can utilize some input from a contractor, often it does not, except in design-build projects. It may be performed by an architect/engineer/contractor firm that possesses both design and construction capabilities.

The successful completion of the feasibility study marks the first of several transition milestones and probably is the most important. This is the starting point for the project by the owner if the project is approved. See Figure 2-2 of the *PMBOK® Guide*—Third Edition. Sometimes the project is not approved, or the feasibility study shows that the project does not meet the financial criteria, and the project is cancelled. It is better to cancel the project at an early stage rather than spend more money on a project that, most likely, will fail ultimately.

2.1.2 Characteristics of Project Phases

See Section 2.1.2 and Figure 2-3 of the *PMBOK® Guide*—Third Edition for an explanation of project phases.

Most construction projects can be viewed in five phases, although they are sometimes shortened to four; each one of these phases can be treated like a project in itself, with all of the process groups operating as they do for the overall project. These phases are concept, planning (and development), detailed design, construction, and start-up and turnover.

The *concept* phase is essentially the feasibility study, which ends with its completion and approval of the project. In the *planning and development* phase, the concept is defined further, the project criteria are established, and basic drawings are produced along with a schedule, budget, and work plan of how the *detailed design, construction,* and *start-up* are to be performed. It is customary, and often critical, that the client or owner approve the basic drawings, criteria, and work plan, which then become the baseline configuration for the project. In the detailed design phase, all design details are completed and drawings and specifications are issued for construction. This can be accomplished using one of the following methods:

- A traditional design-bid-build project delivery method that results in completion of the design work and subsequently, requests for competitive construction bids.
- A design-build or phased design method in which design and construction are performed in parallel. While this method is not new, it is seeing a current surge in popularity. Design is performed in sequential packages, which are then bid and constructed in the order released. This method is sometimes used in an attempt to fast-track a project, but can have negative consequences if an overall design is not locked in early in the project.

When construction is completed, the project is ready for final testing and start-up operations, followed by turnover to the owner. For major projects (e.g., industrial), the start-up phase is often done in sequential segments following the process flow; the project culminates when the plant or project becomes operational according to its design.

The critical milestones for construction projects are:

- Initial approval of the concept (feasibility study)
- Approval of the project criteria (baseline configuration)
- Readiness to initiate start-up
- Contractual completion of the project

Project success depends upon the clear establishment of these milestones and the criteria that defines them.

2.1.3 Project Life Cycle and Product Life Cycle Relationships

See Section 2.1.3 of the *PMBOK® Guide*—Third Edition.

2.2 Project Stakeholders

See Section 2.2 of The *PMBOK® Guide*—Third Edition. In addition to the basic eight stakeholders discussed, every construction project involves one additional stakeholder and many involve a second or third as follows:

- **Regulatory agencies.** Federal, state, local, and international agencies, which issue permits and otherwise control certain aspects of the construction process.
- **General public.** These can be organized groups or citizens who are affected by the construction project during the building process or the facility's operation upon completion of the project. Such public groups have had increasing influence in delaying, modifying the scope of, or, in some cases, causing the cancellation of projects. Refer to Chapter 15 of this extension on Project Financial Management. Public funding sources, such as bonds, tax levies, World Bank, and federal aid, needs to be considered in the financial plan for the project. Funding participants are all very important stakeholders in today's world and need to be included in the analysis of stakeholder needs.
- **Project managers.** Project managers are often assigned to a construction project from organizations, both public and private, that are involved in the construction. The project managers are often changed from one phase to another or some may specialize in only one phase of the construction project.

2.3 Organizational Influences

See Section 2.3 of the *PMBOK® Guide*—Third Edition.

2.3.1 Organizational Systems

See Section 2.3.1 of the *PMBOK® Guide*—Third Edition.

2.3.2 Organizational Cultures and Styles

See Section 2.3.2 of the *PMBOK® Guide*—Third Edition. In construction, the importance of culture may be emphasized when the performing organization's culture differs from that of the owner or customer. This is particularly important to the project when an owner hires the services of an architect/engineer or construction manager to be a part of an integrated project team consisting of employees of the owner and contractor. The melding of different organizational cultures needs to be addressed in order for the project to be successful.

2.3.3 Organizational Structure

The *PMBOK® Guide*—Third Edition has a good discussion on the various types of project organizations but is generally more applicable to the owner's or customer's organizations and overall project organizations. Large engineering firms or engineer/ constructors will likely utilize a matrix organization as shown in Figures 2-9, 2-10, and 2-11 of the *PMBOK® Guide*—Third Edition, while a contractor will be organized more along the lines of a projectized organization (see Figure 2-8 of the *PMBOK® Guide*—Third Edition). The management of construction is sometimes made difficult when several organizational structures intersect because of the various stakeholders' differing viewpoints and agendas. The challenge is to manage the decision process effectively, recognizing these differences. Construction now utilizes partnering as a project alignment technique to bridge these differences. See Section 4.4.2.3.

2.3.4 The Role of the PMO in Organizational Structures

See Section 2.3.4 of the *PMBOK® Guide*—Third Edition.

2.3.5 Project Management System

See Section 2.3.5 of the *PMBOK® Guide*—Third Edition.

Section II

The Standard for Project Management of a Project

Chapter 3

Project Management Processes for a Project

Sections 3.1 and 3.2, of the *PMBOK® Guide*—Third Edition, describe project management processes and lists the five Process Groups, which are also fundamental for a construction project. The interaction between and within each group is outlined in Section 3.3 of the *PMBOK® Guide*—Third Edition. The description of these processes, process groups, and linkages is directly applicable to a construction project, with only a few additions noted in 3.2.2 and 3.2.5 (see Table 3-1 for a complete mapping of the process groups and the Knowledge Areas).

3.1 Project Management Processes

See Section 3.1 of The *PMBOK® Guide*—Third Edition.

3.2 Project Management Process Groups

See Section 3.2 of the *PMBOK® Guide*—Third Edition. Understanding the Process Groups in construction can be a more complex task because there are interrelationships within and among these Process Groups in this industry.

3.2.1 Initiating Process Group

See Section 3.2.1 of the *PMBOK® Guide*—Third Edition.

3.2.2 Planning Process Group

See Section 3.2.2 of the *PMBOK® Guide*—Third Edition. The project management processes outlined in Sections 3.2.2.1 through 3.2.2.21 apply to construction projects. Additional information regarding plan contracting as it applies to construction is described below:

.21 **Plan Contracting**

See Section 3.2.21 of the *PMBOK® Guide*—Third Edition. In the construction industry, determining how to procure the item is just as important as determining what and when. In most procurement activities there are several options available for purchases or subcontracts, such as fixed-price competition, sole source, cost reimbursable, and design-build. Determining which option to use depends on several factors, which are discussed in Chapter 12.

3.2.3 Executing Process Group

See Section 3.2.3 of the *PMBOK® Guide*—Third Edition.

3.2.4 Monitoring and Controlling Process Group

See Section 3.2.4 of the *PMBOK® Guide*—Third Edition.

3.2.5 Closing Process Group

See Section 3.2.5 of the *PMBOK® Guide*—Third Edition. In addition to the processes listed in this section of the *PMBOK® Guide* (Close Project and Contract Closure), closure of the team in a construction project is often a substantial effort; therefore, an additional process has been identified as follows (see also Table 3-1):

.3 **Close Project Team**

This is the activity covering the closeout and dissolution of the project team, primarily at the end of the project. This process is covered in Chapter 9 on Project Human Resource Management.

3.3 Process Interactions

See Section 3.3 of the *PMBOK® Guide*—Third Edition.

3.4 Project Management Process Mapping

See Section 3.4 of the *PMBOK® Guide*—Third Edition.

Process Groups Knowledge Area	Initiating Process Group	Planning Process Group	Executing Process Group	Monitoring and Controlling Process Group	Closing Process Group
4. Project Integration Management	4.1 Develop Project Charter 4.2 Develop Preliminary Project Scope Statement	4.3 Develop Project Management Plan	4.4 Direct and Manage Project Execution	4.5 Monitor and Control Project Work 4.6 Integrated Change Control	4.7 Close Project
5. Project Scope Management		5.1 Scope Planning 5.2 Scope Definition 5.3 Create WBS		5.4 Scope Verification 5.5 Scope Change Control	
6. Project Time Management		6.1 Activity Definition 6.2 Activity Sequencing 6.3 Activity Resource Estimating 6.4 Activity Duration Estimating 6.5 Schedule Development ***6.7 Activity Weights Definitions*** ***6.8 Progress Curves Development***		6.6 Schedule Control ***6.9 Progress Monitoring***	
7. Project Cost Management		7.1 Cost Estimating 7.2 Cost Budgeting		7.3 Cost Control	
8. Project Quality Management		8.1 Quality Planning	8.2 Perform Quality Assurance	8.3 Perform Quality Control	
9. Project Human Resource Management		9.1 Human Resource Planning	9.2 Acquire Project Team 9.3 Develop Project Team	9.4 Manage Project Team	***9.5 Close Project Team***
10. Project Communications Management		10.1 Communications Planning	10.2 Information Distribution	10.3 Performance Reporting 10.4 Manage Stakeholders	
11. Project Management Risk		11.1 Risk Management Planning 11.2 Risk Identification 11.3 Qualitative Risk Analysis 11.4 Quantitative Risk Analysis 11.5 Risk Response Planning		11.6 Risk Monitoring and Control	
12. Project Procurement Management		12.1 Plan Purchases and Acquisitions 12.2 Plan Contracting	12.3 Request Seller Responses 12.4 Select Sellers	12.5 Contract Administration	12.6 Contract Closure
13. Project Safety Management		***13.1 Safety Planning***	***13.2 Perform Safety Assurance***	***13.3 Perform Safety Control***	
14. Project Environmental Management		***14.1 Environmental Planning***	***14.2 Perform Environmental Assurance***	***14.3 Perform Environmental Control***	
15. Project Financial Management		***15.1 Financial Planning***		***15.2 Perform Financial Control***	***15.3 Perform Financial Administration and Records***
16. Project Claim Management		***16.1 Claim Identification*** ***16.2 Claim Quantification***		***16.3 Claim Prevention***	***16.4 Claim Resolution***

Table 3-1. Mapping of the Project Management Processes to the Project Management Process Groups and the Knowledge Areas

Section III

The Project Management Knowledge Areas

Chapter 4

Project Integration Management

According to the *PMBOK® Guide*—Third Edition, "The Project Integration Management Knowledge Area includes the processes and activities needed to identify, define, combine, unify, and coordinate the various processes and activities of project management within the Project Process Groups." Chapter 4 of the *PMBOK® Guide*—Third Edition describes the following seven major processes:

4.1 Develop Project Charter
4.2 Develop Preliminary Project Scope Statement
4.3 Develop Project Management Plan
4.4 Direct and Manage Project Execution
4.5 Monitor and Control Project Work
4.6 Integrated Change Control
4.7 Close Project

All of these processes apply to projects in the construction industry, with only slight additions or modifications. The need to integrate all elements integrated and quickly reflect any changes in the project plan as it is executed is particularly important in construction. In addition, the design and construction industry maintains the ability to work across borders on a global basis; therefore it must adhere to the jurisdictional regulations of the residing government. This industry also has many unique standards and disciplines that must be fully integrated in support of the project, such as financial considerations of the stakeholders, safety, security, environmental regulations, and the always-present potential for legal disputes and consequences. Refer to Chapters 13, 14, 15, and 16 of this extension. Further, the requirement for complete project documentation, quality and performance records, and the historical records of the transpired communications, enforces the need for a fully integrated project plan and all of the processes included in the *PMBOK® Guide*—Third Edition. See Figure 4-1 for an overview of the processes and respective inputs, tools and techniques, and outputs.

PROJECT INTEGRATION MANAGEMENT

4.1 Develop Project Charter

.1 Inputs
- .1 Contract (when applicable)
- .2 Project statement of work
- .3 Enterprise environmental factors
- *.4 Organizational process assets*

.2 Tools and Techniques
- .1 Project selection methods
- .2 Project management methodology
- .3 Project management information system
- .4 Expert judgment

.3 Outputs
- .1 Project charter

4.2 Develop Preliminary Project Scope Statement

.1 Inputs
- .1 Project charter
- .2 Project statement of work
- .3 Enterprise environmental factors
- .4 Organizational process assets
- ***.5 Contract documents***

.2 Tools and Techniques
- .1 Project management methodology
- .2 Project management information system
- .3 Expert judgment

.3 Outputs
- .1 Preliminary project scope statement

4.3 Develop Project Management Plan

.1 Inputs
- .1 Preliminary project scope statement
- .2 Project management processes
- .3 Enterprise environmental factors
- .4 Organizational process assets
- ***.5 Contract documents***

.2 Tools and Techniques
- .1 Project management methodology
- .2 Project management information system
- .3 Expert judgment
- ***.4 Company procedures***
- ***.5 Value engineering***

.3 Outputs
- .1 Project management plan
- ***.2 Project management strategy***

4.4 Direct and Manage Project Execution

.1 Inputs
- .1 Project management plan
- .2 Approved corrective actions
- .3 Approved preventive actions
- .4 Approved change requests
- .5 Approved defect repair
- .6 Validated defect repair
- *.7 Administrative closure procedure*

.2 Tools and Techniques
- *.1 Project management methodology*
- .2 Project management information system
- ***.3 Partnering***
- ***.4 Value engineering***

.3 Outputs
- .1 Deliverables
- .2 Requested changes
- .3 Implemented change requests
- .4 Implemented corrective actions
- .5 Implemented preventive actions
- .6 Implemented defect repair
- .7 Work performance information
- ***.8 Project management strategy***

4.5 Monitor and Control Project Work

.1 Inputs
- .1 Project management plan
- .2 Work performance information
- .3 Rejected change requests

.2 Tools and Techniques
- .1 Project management methodology
- .2 Project management information system
- .3 Earned value technique
- .4 Expert judgment
- ***.5 Partnering follow-up and evaluations***

.3 Outputs
- .1 Recommended corrective actions
- .2 Recommended preventive actions
- .3 Forecasts
- .4 Recommended defect repair
- .5 Requested changes

4.6 Integrated Change Control

.1 Inputs
- .1 Project management plan
- .2 Requested changes
- .3 Work performance information
- .4 Recommended preventive actions
- .5 Recommended corrective actions
- .6 Recommended defect repair
- .7 Deliverables

.2 Tools and Techniques
- .1 Project management methodology
- .2 Project management information system
- .3 Expert judgment

.3 Outputs
- .1 Approved change requests
- .2 Rejected change requests
- .3 Project management plan (updates)
- .4 Project scope statement (updates)
- .5 Approved corrective actions
- .6 Approved preventive actions
- .7 Approved defect repair
- .8 Validated defect repair
- .9 Deliverables

4.7 Close Project

.1 Inputs
- .1 Project management plan
- *.2 Contract documentation*
- .3 Enterprise environmental factors
- .4 Organizational process assets
- .5 Work performance information
- .6 Deliverables

.2 Tools and Techniques
- .1 Project management methodology
- .2 Project management information system
- .3 Expert judgment
- ***.4 Develop project punchlist***

.3 Outputs
- *.1 Administrative closure procedure*
- .2 Contract closure procedure
- .3 Final product, service, or result
- .4 Organizational process assets (updates)
- ***.5 Lessons learned***

Figure 4-1. Project Integration Management Overview

4.1 Develop Project Charter

See Section 4.1 of the *PMBOK® Guide*—Third Edition.

4.1.1 Develop Project Charter: Inputs

See Sections 4.1.1.1 through 4.1.1.3 of the *PMBOK® Guide*—Third Edition, with the modification of 4.1.1.4 as follows:

.4 Organizational Process Assets

See Section 4.1.1.4 of the *PMBOK® Guide*—Third Edition. Design and construction projects are exposed to physical and environmental risks in greater measure than other parts of the project. These particular risks are part of and should be included in the total of all project risks. The organizational process assets should establish means and methods for integrating the safety, environmental, and security risks of the project. See also Chapters 13 and 14 of this extension. Each construction site, as part of its project plan, needs to have established special plans to address environmental consequences, personnel and public safety, and security considerations. Such items would include theft prevention, damage control for hazardous materials, and secure work zones often mandated for airports, correctional institutions, or private businesses.

4.1.2 Develop Project Charter: Tools and Techniques

See Section 4.1.2 of the *PMBOK® Guide*—Third Edition.

4.1.3 Develop Project Charter: Outputs

See Section 4.1.3 of the *PMBOK® Guide*—Third Edition.

4.2 Develop Preliminary Project Scope Statement

See Section 4.2 of the *PMBOK® Guide*—Third Edition. For design and construction projects, the statement of requirements, which provides details about specific project requirements, is necessary for effective resource planning. It should include details about the site, outline of design parameters, outline of engineering requirements, technical definitions, construction timescale, applicable codes and standards, inspection/testing requirements, safety-related requirements, and spare policy.

4.2.1 Develop Preliminary Project Scope Statement: Inputs

See Section 4.2.1 of the *PMBOK® Guide*—Third Edition. The construction industry is not limited to the geographical borders of an organization, but it is, however, constrained by the regulations and jurisdictional laws of the location where the construction is performed. Most organizations performing work within a geographical location or internationally, educate themselves on both the contractual and the jurisdictional regulations. One vitally important input that is essential for these traveling firms is the enterprise environmental factors. Firms performing work must adhere to, with strict conformance, the laws and governmental regulations that affect the project, its locations, its labor, and environmental restrictions. Items of particular importance are:

- Government or industry standards (e.g., regulatory agency regulations, product standards, quality standards, and workmanship standards);
- Infrastructure requirements (e.g., existing and temporary facilities for construction);
- Existing and required human resources (e. g., as it pertains to construction, skills, disciplines, and knowledge such as design, development, legal, contracting, and purchasing), and
- Personnel administration (e.g., hiring and firing guidelines, employee compensation, imported labor, and security records).

Sections 4.2.1.1 through 4.2.1.4 of the *PMBOK® Guide*—Third Edition discuss inputs from the Develop Preliminary Project Scope Statement process. An additional input specific to construction projects is described as follows:

.5 Contract Documents

Design and construction projects incorporate and define the project requirements in addition to the constraints and protocols of how the project needs to be constructed. Contract documents include the contract and conditions which outline the manner in which the project is administered. Other contract documents include applicable reference documents and standards for codes, procedures and processes and the technical specifications, which outline the materials and installation processes. See Plan Contracting (Section 12.2), which describes the process which integrates the project contract documents by means of subcontracts and purchase orders for material, equipment and specialty contractors.

4.2.2 Develop Preliminary Project Scope Statement: Tools and Techniques

See Section 4.2.2 of the *PMBOK® Guide*—Third Edition.

4.2.3 Develop Preliminary Project Scope Statement: Outputs

See Section 4.2.3 of the *PMBOK® Guide*—Third Edition.

4.3 Develop Project Management Plan

See Section 4.3 of the *PMBOK® Guide*—Third Edition. Virtually all construction projects are carried out under the terms of a written contract between the owner and the contractor(s). In some cases the contract can be considered an input to project plan development and, in others, it may be the output. For the purposes of this section, it will be considered to be an input. Consequently, the contract for these terms provides an overall constraint or boundary for the development of the project plan and a foundation of a portion of the plan itself.

4.3.1 Develop Project Management Plan: Inputs

See Section 4.3.1 through 4.3.1.4 of the *PMBOK® Guide*—Third Edition. See also Section 4.3.1.5 on Contract Documents, which is a significant input to the project management plan.

.5 Contract Documents

The inputs to developing the project management plan include enterprise environmental factors. In addition, there are major factors that apply to construction in foreign countries by contractors that are not residents of the host country: (a) Export controls on materials—the project manager must consider import/ export regulations of both the host country and exporting country; (b) Security—consideration must be given to security requirements at the work-side residencies for workers offsite, and transportation routes for materials; and (c) Cultural norms—Contractor employees who are not residents of the host country may need sensitivity training to avoid inadvertent incidents where cultural values and customs may be violated which could lead to unrest. Customs of the countries and people involved must be respected. Construction projects may also have requirements for operator training or other special requirements called for by the contract. Special attention should be made to incorporate, either by reference or in full detail, the many special plans required for construction projects. Detailed plans such as safety, quality security, environmental, and human resource plans are all project-specific and must be integrated in order to fully comply with the contract requirements.

4.3.2 Develop Project Management Plan: Tools and Techniques

See Sections 4.3.2.1 through 4.3.2.3 of the *PMBOK® Guide*—Third Edition. For the construction industry, two new items, company procedures and value engineering, have been added here as follows:

.4 Company Procedures

Many companies, engineer/contractors, and contractors, have a standard set of policies and procedures that guide a company's approach to developing a project plan. While many of the aspects of this process can make good use of the company's standard procedures, it may be necessary to develop a distinct plan for certain projects.

.5 Value Engineering

Value engineering is particularly useful in developing the project plan and scope. United States Office of Management and Budget (OMB) Circular A-131 defines value engineering as "an organized effort directed at analyzing the functions of systems, equipment, facilities, services, and supplies for the purpose of achieving the essential functions at the lowest life cycle cost consistent with required, quality, reliability, and safety." As such, whether formal or informal, this process can have a major impact on the success of a project.

This process, which seeks the best or optimal way of accomplishing many of the activities of a project, is a useful tool in developing the project plan. While its main use may be in defining the scope (see Section 5.1.2), the process can also be applied to a host of other activities. For example, in a closely sited industrial plant with restricted access, it may ultimately be less costly to use underground conduit runs so roads can be opened for the balance of construction rather than use overhead tray and conduit, which can block large equipment access. Alternatives developed must be monitored for incorporation or rejection throughout the project.

Similar to the value engineering process, is the Charrette workshop, which often takes place at the conception stage of a project. Originating as a French architectural programming process, it can be utilized, to bring together the stakeholders and designers into working sessions geared towards developing an optimal design that meets the functional needs of the facility user, the design aesthetics for the appearance and style of the constructed product and economical constraints of the stakeholders. While the structure of a Charrette varies depending on the design problem and the individuals in the group, Charrettes often take place in the form of a workshop and may include multiple sessions. Refer to Chapter 5 on Project Scope Management of this extension, specifically, Section 5.2.2.2 on Alternatives Identification.

4.3.3 Develop Project Management Plan: Outputs

See Section 4.3.3 of the *PMBOK® Guide*—Third Edition, with the addition of 4.3.3.2 as follows:

.2 Project Management Strategy

An important output for the Direct and Manage Project Execution process is the development of a project management strategy which will be dictated by all preceding project planning processes. The project management strategy is fundamentally different from a performing organization's project management policy which details the overall intentions of the performing organization. The Project management strategy will include but are not limited to the following:

- How the project management team will react to changes in influencing factors which are often seen as minor issues at project commencement become more prevalent as the project advances.
- Provision of human resources, i.e. external or internal
- Methods for procurement and subcontract
- Methods for project stakeholder management

The project management strategy would also include the approaches the performing organization will take towards the management and execution of the other knowledge areas such as project quality management (See Chapter 8 of the *PMBOK® Guide*—Third Edition and Chapter 8 of this extension), Project Safety Management (See Chapter 13 of this extension), etc. The project management strategy needs to be carefully monitored to ensure that in remains pertinent to the work practices adopted (and vice versa), and, equally important, that it remains applicable to the prevailing conditions (economic, social, etc.) which can change as a project progresses. It is therefore seen as a key deliverable for any construction project.

4.4 Direct and Manage Project Execution

See Section 4.4 of the *PMBOK® Guide*—Third Edition.

4.4.1 Direct and Manage Project Execution: Inputs

See Section 4.4.1 of the *PMBOK® Guide*—Third Edition. See Section 4.4.1 of the *PMBOK® Guide*—Third Edition, with the modification of 4.4.1.7, as follows:

.7 Administrative Closure Procedure

Specific to the construction industry are the physical and administrative procedures for commissioning (start-up) of the completed facility. Often, these work performance constraints go hand-in-hand with the construction process to ensure that the quality, performance, and longevity of the facility meet the design intent. For contracts with specific requirements, a commissioning plan should be made a part of the project plan. To comply with the jurisdiction codes and compliance aspects of the project, specific aspects of the project must be inspected for compliance with environmental, health, and life safety elements in order to receive an approval for beneficial occupancy, which is a required administrative closure requirement. See also Chapter 3 of the *PMBOK® Guide*—Third Edition, for the life cycle of projects for various types of industries and specifically for construction.

4.4.2 Direct and Manage Project Execution: Tools and Techniques

See Sections 4.4.2.1 and 4.4.2.2 of the *PMBOK® Guide*—Third Edition. Refer to the specific construction-related modification in Section 4.4.3.1 of this extension, in addition to the new tools and techniques provided in Sections 4.4.3.3 and 4.4.3.4 as follows:

.1 Project Management Methodology

See also 4.4.2.1 of the *PMBOK® Guide*—Third Edition. In construction, the quality monitoring and control plan is a detailed system that describes the inspection process, including the frequency, level of inspection, component testing, and performance reviews to be initiated on a project. Its purpose is consistent with the intent of the design, which is to provide adequate details and descriptions for the construction process and to ensure compliance with the many building safety and environmental codes. To ensure that the work performance, along with the blended use of materials with workmanship meet the design intent, the contract documents often detail the sometimes-rigorous inspection and verification of the permanent materials, fabrications and on-site constructed components. See Chapter 8 on Project Quality Management, for additional information on the inspection and quality programs within the construction industry. The reporting system must capture and communicate the results of inspections. See Section 10.2 Information Distribution. The results of these inspections become an output within Section 4.4.3.7 on Work Performance Information.

.3 Partnering

Partnering is a process that creates a personal commitment among the project participants. These commitments transcend the contract and project requirements to create an environment where the participants recognize the personal value of successfully completing the projects in spite of the often difficult and frustrating circumstances encountered. The technique of creating a team environment through partnering can complement even the best efforts of the individual organizations involved with the execution and administration of the construction project. Known for its project management alignment fundamentals, partnering can immediately create a project environment of collaboration and teamwork. See Section 16.3.2.5 of this extension on Partnering.

.4 Value Engineering

See also Section 4.3.2.5 of this extension. Value engineering is a process that creates a shared incentive between project participants. The process is intended to lower the project costs, shorten work durations, or increase production without decreasing the product quality, by altering the means and methods of the design and construction applications, without diverting the purpose, objective, or goals of the customer.

4.4.3 Direct and Manage Project Execution: Outputs

See Section 4.4.3 of the *PMBOK® Guide*—Third Edition, with the addition of Section 4.4.3.8 on project management strategy as follows:

.8 Project Management Strategy

An important output for the Direct and Manage Project Execution process is the development of a project management strategy which will be dictated by all preceding project planning processes. The project management strategy is fundamentally different from a performing organization's project management policy which details the overall intentions of the performing organization. The project management strategy will include but is not limited to the following:

- How the project management team will react to changes in influencing factors which are often seen as minor issues when project commencement becomes more prevalent as the project advances.
- Provision of human resources, that is, external or internal.
- Methods for procurement and subcontract.
- Methods for project stakeholder management.

The project management strategy would also include the approaches that the performing organization will take towards the management and execution of the other Knowledge Areas, such as Project Quality Management (see Chapter 8 of the *PMBOK®Guide*—Third Edition and Chapter 8 of this Extension). The project management strategy needs to be carefully monitored to ensure that it remains pertinent to the work practices adopted (and vice versa), and equally important, that it remains applicable to the prevailing conditions (economic, social, etc.) which can change as a project progresses. It is therefore seen as a key deliverable for any construction project.

4.5 Monitor and Control Project Work

See Section 4.5 of the *PMBOK® Guide*—Third Edition.

4.5.1 Monitor and Control Project Work: Inputs

See Section 4.5.1 of the *PMBOK® Guide*—Third Edition.

4.5.2 Monitor and Control Project Work: Tools and Techniques

See Section 4.5.2 of the *PMBOK® Guide*—Third Edition, with the addition of Section 4.5.2.5 as follows:

.5 Partnering Follow-up and Evaluations

If partnering has been established with the project team, regular on-site follow-up meetings are essential to maintain the team working environment. These follow-up sessions can also serve the purpose of problem solving and dispute resolution within the team and the stakeholders. Other benefits from these regular follow-ups include process and system improvements for timely turnaround of key submittals or decisions on changes. Along with the follow-up meetings, the teams can also evaluate their project management culture and progress towards goals and objectives through the use of a partnering assessment tool. This survey type tool can be customized to reflect the project team's stated goals and working relationship. See Section 16.3 of this extension for further information on claim prevention.

4.5.3 Monitor and Control Project Work: Outputs

See Section 4.5.3 of the *PMBOK® Guide*—Third Edition.

4.6 Integrated Change Control

See Section 4.6 of the *PMBOK® Guide*—Third Edition. One of the most important aspects of plan execution in construction is the control of changes to the project. It is the task of integrated change control as described in this section of the *PMBOK® Guide*—Third Edition to identify possible changes; review them for their effect on project scope, cost, and schedule; notify the owner in accordance with procedures and time requirements, process the change as stated in the contract documents, and ensure that a proper project record is made of the disposition for the change. There are a number of change control and configuration management systems that perform this function, but every construction project must have one that has proven to be effective. An absent or deficient change control system can often produce the most negative effect on a construction project and the reputation of its contractor.

In construction, ultimate control or approval of changes is usually the responsibility of the owner, who is often the source of changes to the project. Changes may occur for a variety of reasons and they often fall in one of these categories: scope change, errors and omissions, and unforeseen conditions. It is important that changes be approved by an authorized representative or authority of the owner. Further, situations may arise in which contractors have acted on a change request by someone not authorized to do so. This may result as a contractor-incurred cost which may not gain owner approval. The contractors may nevertheless initiate a claim to seek compensation. It is the responsibility of the engineer/contractor to identify changes in a timely manner and to advise the owner of their effect on the quality, cost, and time of performance of the project. In larger projects and in some public projects, as stated in the *PMBOK® Guide*—Third Edition, there may be a more formal control board that performs the analysis and renders the approval or rejection of changes on behalf of the owner. It is vital that changes and their effect be reviewed periodically, usually no less than on a monthly basis. There are different layers of changes on a typical construction project from subcontracts, purchase orders, and other agreements that may or may not relate to changes to the project's contracting authority.

4.6.1 Integrated Change Control: Inputs

See Section 4.6.1 of the *PMBOK® Guide*—Third Edition.

4.6.2 Integrated Change Control: Tools and Techniques

See Section 4.6.2 of the *PMBOK® Guide*—Third Edition.

4.6.3 Integrated Change Control: Outputs

See Section 4.6.3 of the *PMBOK® Guide*—Third Edition.

4.7 Close Project

See Section 4.7 of the *PMBOK® Guide*—Third Edition. An important phase in the life of a construction project is administrative closure, whether it comes after either achieving its objectives or being terminated for other reasons. The *PMBOK® Guide*—Third Edition states, "The Close Project process involves performing the project closure portion of the project management plan. In multi-phase projects, the Close Project process closes out the portion of the project scope and associated activities applicable to a given phase. This process includes finalizing all activities completed across all Project Management Process Groups to formally close the project or a project phase, and transfer the completed or cancelled project as appropriate."

Closure activities should not be delayed until project completion; this is a particularly important point in connection with construction projects with the varied and large number of elements and typical lack of funding and personnel to complete the closure process. Each phase of the project should be properly closed to ensure that important and useful information is not lost. Critical among this information are as-built records of the construction showing actual dimensions and elevations of the completed work, especially with regard to underground work, which may need to be repaired or modified later. Industrial projects are normally completed in order of the product process, which makes it possible and beneficial to complete and close portions of the project as the entire project moves to closeout. Also in this process, owners may begin occupying and running completed portions before the entire project is completed (known as beneficial occupancy).

4.7.1 Close Project: Inputs

See Section 4.7.1 of the *PMBOK® Guide*—Third Edition, with the following modification of Section 4.7.1.2 as follows:

.2 Contract Documentation

See also Section 4.7.1.2 of the *PMBOK® Guide*—Third Edition. For construction projects, these documents also include inspection and testing records and reports, operation and maintenance manuals, and similar records that are relevant to the completion and performance of the project. Substantial completion is often a contractual requirement, which ties in directly with the contract time for completion of the project. At this juncture, a final punch list is generated and a time limit is imposed on the contractor to complete all contract and any required rework.

4.7.2 Close Project: Tools and Techniques

See Section 4.7.2 of the *PMBOK® Guide*—Third Edition, with the addition of Section 4.7.2.4 as follows:

.4 Develop Project Punch list

The technique of generating a list of all outstanding contract performance items is known as the punch list. This is sometimes preceded by a preliminary punch list, which allows the contractor and its subcontractors to correct any performance deficiencies prior to the owner or stakeholder occupying any portions of the facility. The prime contractor, the designer and its technical consultants, and various members of the owner/stakeholder organization jointly perform the final punch list walkthrough. This final punch list becomes one of the last closure documents and often initiates the start of various warranty periods for the facility and its component equipment. On process facilities, a significant portion of the punch list may be the operational or performance testing of the completed facility. Refer to Sections 8.2 and 8.3 of the *PMBOK® Guide*—Third Edition for quality management processes. It is also during this period that the various equipment suppliers will conduct operational training for the owner's maintenance and facility members. Refer to Sections 12.2 and 12.6 of this extension for procurement requirements. See also Section 5.4.3 of the *PMBOK® Guide*—Third Edition for scope verification outputs.

4.7.3 Close Project: Outputs

See Section 4.7.3 of the *PMBOK® Guide*—Third Edition, with the modification of Section 4.7.3.1 and the additional output of Section 4.7.3.5, as follows:

.1 Administrative Closure Procedure

See also Section 4.7.3.1 of the *PMBOK® Guide*—Third Edition. Project closure may be realized sequentially as significant portions of the project are completed and turned over to the owner for beneficial occupancy or provisional acceptance. In addition to the formal acceptance documents that are signed by the customer, many construction projects have other documentation required by governmental agencies to be prepared, executed, and distributed. The formal action of final acceptance and closure will in almost all cases be guided by the provision of contractual documents under which the project was constructed.

.5 Lessons Learned

In some cases, owners who anticipate future requirements will ask that a final report of the project be prepared, which describes and documents the history of the project, including what went well and what did not. Lessons learned is the product of a collaborative process among project participants to identify successes and areas for improvement in subsequent projects. Project success, amplified with customer feedback and testimonials, will describe proper execution of elements in the project and the management framework, and cite these as examples that can be repeated. The areas of improvement will describe the hardships or problems, the corrective action undertaken, and the preventative actions to be implemented on future work. The lessons learned output is normally a written document that is developed by and shared with the current and future project participants. In the event that the owner does not want or require such a report, the contractor should prepare the report for assistance with future projects. Tools that can be used to gather information for a project history include project de-briefing sessions with the various participating entities, project evaluation forms, and interviews of the principal participants by independent third parties.

Chapter 5

Project Scope Management

See the introduction to Chapter 5 of the *PMBOK® Guide*—Third Edition. See Figure 5-1 for an overview of the processes and respective inputs, tools and techniques, and outputs.

5.1 Scope Planning

See Section 5.1 of the *PMBOK® Guide*—Third Edition, with the following addition to better express scope planning as it pertains to the construction industry:

For a construction project to be successful, project scope planning should involve all the key players at all levels, the owner, the consultant, the general contractor, subcontractors, and suppliers. Although each will only be involved in their respective areas, success increases with interactive involvement. Value engineering (Section 4.3.2.5) and product analysis (Section 5.1.2.3) can be most useful in scope planning as a tool and technique to obtain an optimal result.

5.1.1 Scope Planning: Inputs

See 5.1.1 of the *PMBOK® Guide*—Third Edition, with the modification of Section 5.1.1.3, and the additions of Sections 5.1.1.6, 5.1.1.7, and 5.1.1.8 as follows:

.3 Project Charter

See also Section 5.1.1.3 of the *PMBOK® Guide*—Third Edition, which refers to Section 4.1, which makes reference to contract provisions. In fact, the contract may define a large part of the scope (depending on what phase of the construction project it is issued in) and certainly forms an important part of the input to scope planning.

.6 Contract Documents

See Section 4.7.1.2 of the *PMBOK® Guide*—Third Edition on Contract Documentation of the *PMBOK® Guide*—Third Edition. The primary input document for construction projects is the contract, which contains the scope of work for the contractor to perform. The scope is typically described in customer technical specifications, drawings, legal terms and conditions, and various other technical and administrative requirements of the project.

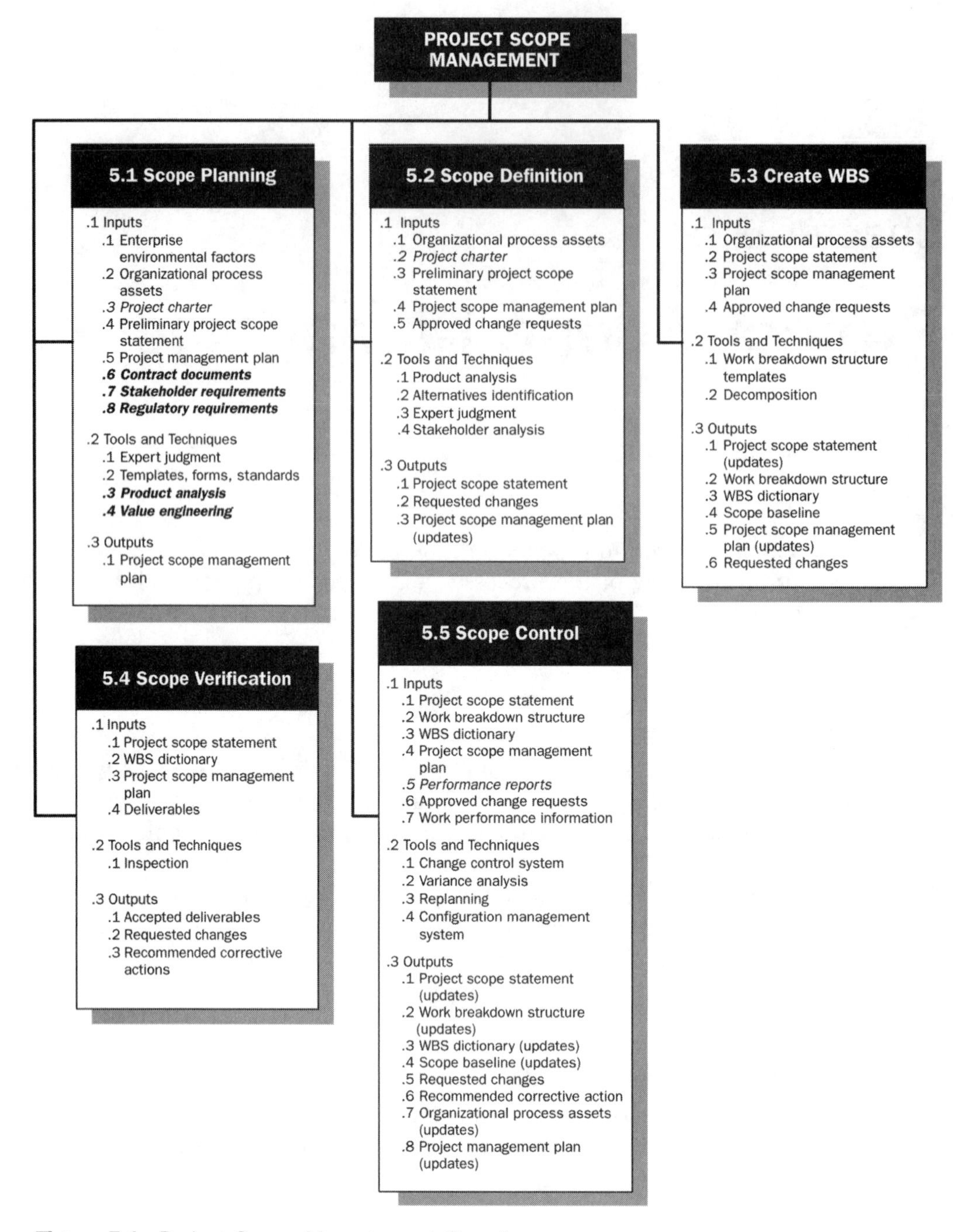

Figure 5-1. Project Scope Management Overview

.7 Stakeholder Requirements

See Section 2.2 of the *PMBOK® Guide*—Third Edition on Project Stakeholders of the *PMBOK® Guide*—Third Edition. The requirements imposed by stakeholders can influence the scope of the project as well as the performance of the work to be performed.

.8 Regulatory Requirements

To project the safety and health of construction personnel and the public, construction projects are normally required to meet minimum requirements established by the government. In addition, the government often requires a construction project to receive a "statutory" permit prior to commencing construction. The permit may specify requirements that impact the project scope.

5.1.2 Scope Planning: Tools and Techniques

See Section 5.1.2 of the *PMBOK® Guide*—Third Edition. Additional requirements for scope planning, as it pertains to construction projects, are included in 5.1.2.3 and 5.1.2.4.

.3 Product Analysis

During the development of the facility (or product), product analysis can be used to review several options for design, ways of accomplishing work, and alternatives to achieve the essential functions of the facility at the lowest life cycle cost consistent with other important parameters (see Section 4.3.2.5 and Section 5.1.2.4 in this extension).

.4 Value Engineering

One of the major uses of value engineering is to examine the alternative ways of providing the functions of a facility and to provide a means for determining which of these alternatives furnish the optimum result. Under the formal system of value engineering, a team of engineers may also review such things as; improving productivity, simplifying work, conserving energy and water, and re-evaluating service contracts. (SAVE International is an organization dedicated to the application and standardization of the value engineering process and can provide additional information on the subject.) In some cases, contractors may furnish better and more economical ways of accomplishing work if they are allowed to submit alternate bids. This is discussed further in Chapter 12 of this extension.

5.1.3 Scope Planning: Outputs

See Section 5.1.3 of the *PMBOK® Guide*—Third Edition.

5.2 Scope Definition

See Section 5.2 of the *PMBOK® Guide*—Third Edition.

5.2.1 Scope Definition: Inputs

See Section 5.2.1 of the *PMBOK® Guide*—Third Edition, with the modification of Section 5.2.1.2 as follows:

.2 Project Charter

See Section 5.2.1.2 of the *PMBOK® Guide*—Third Edition. All construction projects are performed under some form of contract, therefore, it is important to stress the clarity of the contract language to avoid or minimize misunderstandings or errors in its interpretation.

5.2.2 Scope Definition: Tools and Techniques

See Section 5.2.2 of the *PMBOK® Guide*—Third Edition

5.2.3 Scope Definition: Outputs

See Section 5.2.3 of the *PMBOK® Guide*—Third Edition.

5.3 Create WBS

See Section 5.3 of the *PMBOK® Guide*—Third Edition. The PMI Practice Standard for Work Breakdown Structures provides additional information in the creation of work breakdown structures.

5.4 Scope Verification

See Section 5.4 of the *PMBOK® Guide*—Third Edition, which defines scope verification as "the process of obtaining the stakeholders' formal acceptance of the completed project scope and associated deliverables. Verifying the project scope includes reviewing deliverables and work results to ensure that each is completed satisfactorily."

While this definition describes the ultimate acceptance of the work by the owner or authorizing party, construction projects are conducted in clearly defined phases and there are other verification steps along the way that are required. The first phase is at the end of the concept phase when the project is approved. The contractor may or may not be involved in this process but it results in a preliminary scope and most likely a contract generally outlining what is to be constructed. The next phase is a definition phase, when sufficient plans and specifications have been developed to provide a baseline criteria, budget, and schedule. The final phase, the acceptance of the project, is described in this section of the *PMBOK® Guide*—Third Edition as being properly completed in accordance with the contract. Completion of each of these steps should be marked by a formal verification process before proceeding to the next step. The processes listed in this section can be used to obtain this verification also.

5.4.1 Scope Verification: Inputs

See Section 5.4.1 of the *PMBOK® Guide*—Third Edition.

5.4.2 Scope Verification: Tools and Techniques

See Section 5.4.2 of the *PMBOK® Guide*—Third Edition.

5.4.3 Scope Verification: Outputs

See Section 5.4.3 of the *PMBOK® Guide*—Third Edition.

5.5 Scope Control

See Section 5.5 of the *PMBOK® Guide*—Third Edition.

5.5.1 Scope Control: Inputs

See Section 5.5.1 *PMBOK® Guide*—Third Edition. See also Sections 5.5.1.1 through 5.5.1.7, with the additional construction-related information for Section 5.5.1.5 as follows:

.5 Performance Reports

See also Section 5.5.1.5 of the *PMBOK® Guide*—Third Edition. In most construction projects, performance, that is whether or not the project is on schedule and on budget, does not affect the scope. However, in some cases under a reimbursable contract, a budget overrun might result in a necessary reduction in scope. Similarly, an underrun might allow the owner to add features that were not included in the original scope.

5.5.2 Scope Control: Tools and Techniques

See Section 5.5.2 of the *PMBOK® Guide*—Third Edition.

5.5.3 Scope Control: Outputs

See Section 5.5.3 of the *PMBOK® Guide*—Third Edition.

Chapter 6

Project Time Management

See the introduction to the *PMBOK® Guide*—Third Edition, which describes six processes under Project Time Management. The PMI Practice Standard for Scheduling provides additional information to aid in the development of project schedules.

6.1 Activity Definition.
6.2 Activity Sequencing.
6.3 Activity Resource Estimating.
6.4 Activity Duration Estimating.
6.5 Schedule Development.
6.6 Schedule Control.

For construction, three additional processes are also usually required as follows:

6.7 Activity Weights Definition—determining the relative and absolute weights for each project activity.
6.8 Progress Curves Development—analyzing activity weights and project schedule to create progress curves.
6.9 Progress Monitoring—monitoring project progress.

See Figure 6-1 for an overview of the processes and respective inputs, tools and techniques, and outputs.

6.1 Activity Definition

See Section 6.1 of the *PMBOK® Guide*—Third Edition.

6.1.1 Activity Definition: Inputs

See Sections 6.1.1 of the *PMBOK® Guide*—Third Edition, with the modification of Section 6.1.1.4 and the addition of Section 6.1.1.7 as follows:

.4 Work Breakdown Structure

The WBS used in the scheduling process must be consistent with those of the cost and human resource management processes so that integration between schedule, cost, and responsibilities can be achieved (see Section 5.3 of the *PMBOK® Guide*—Third Edition for a more detailed discussion of the WBS).

PROJECT TIME MANAGEMENT

6.1 Activity Definition

.1 Inputs
- .1 Enterprise environmental factors
- .2 Organizational process assets
- .3 Project scope statement
- *.4 Work breakdown structure*
- .5 WBS dictionary
- .6 Project management plan
- ***.7 Constraints***

.2 Tools and Techniques
- *.1 Decomposition*
- .2 Templates
- .3 Rolling wave planning
- .4 Expert judgment
- .5 Planning component
- ***.6 Concurrent engineering***

.3 Outputs
- .1 Activity list
- .2 Activity attributes
- .3 Milestone list
- .4 Requested changes

6.2 Activity Sequencing

.1 Inputs
- .1 Project scope statement
- .2 Activity list
- .3 Activity attributes
- .4 Milestone list
- .5 Approved change requests

.2 Tools and Techniques
- .1 Precedence Diagramming Method (PDM)
- .2 Arrow Diagramming Method (ADM)
- .3 Schedule network templates
- .4 Dependency determination
- .5 Applying leads and lags
- ***.6 Linear scheduling method***

.3 Outputs
- .1 Project schedule network diagrams
- .2 Activity list (updates)
- .3 Activity attributes (updates)
- .4 Requested changes

6.3 Activity Resource Estimating

.1 Inputs
- .1 Enterprise environmental factors
- .2 Organizational process assets
- .3 Activity list
- .4 Activity attributes
- .5 Resource availability
- .6 Project management plan

.2 Tools and Techniques
- .1 Expert judgment
- .2 Alternatives analysis
- .3 Published estimating data
- .4 Project management software
- .5 Bottom-up estimating

.3 Outputs
- .1 Activity resource requirements
- .2 Activity attributes (updates)
- .3 Resource breakdown structure
- .4 Resource calendars (updates)
- .5 Requested changes

6.4 Activity Duration Estimating

.1 Inputs
- .1 Enterprise environmental factors
- .2 Organizational process assets
- .3 Project scope statement
- .4 Activity list
- .5 Activity attributes
- .6 Activity resource requirements
- .7 Resource calendars
- .8 Project management plan
  - Risk register
  - Activity cost estimates

.2 Tools and Techniques
- .1 Expert judgment
- .2 Analogous estimating
- .3 Parametric estimating
- .4 Three-point estimates
- .5 Reserve analysis

.3 Outputs
- .1 Activity duration estimates
- .2 Activity attributes (updates)

6.5 Schedule Development

.1 Inputs
- .1 Organizational process assets
- .2 Project scope statement
- .3 Activity list
- .4 Activity attributes
- .5 Project schedule network diagrams
- .6 Activity resource requirements
- .7 Resource calendars
- .8 Activity duration estimates
- .9 Project management plan
  - Risk register
- ***.10 Constraints***

.2 Tools and Techniques
- .1 Schedule network analysis
- .2 Critical path method
- .3 Schedule compression
- .4 What-if scenario analysis
- .5 Resource leveling
- .6 Critical chain method
- .7 Project management software
- .8 Applying calendars
- .9 Adjusting leads and lags
- .10 Schedule model
- ***.11 Linear scheduling method***

.3 Outputs
- .1 Project schedule
- .2 Schedule model data
- .3 Schedule baseline
- .4 Resource requirements (updates)
- .5 Activity attributes (updates)
- .6 Project calendar (updates)
- .7 Requested changes
- .8 Project management plan (updates)
  - Schedule management plan (updates)

6.6 Schedule Control

.1 Inputs
- .1 Schedule management plan
- .2 Schedule baseline
- .3 Performance reports
- .4 Approved change requests
- ***.5 Schedule impacts***

.2 Tools and Techniques
- .1 Progress reporting
- .2 Schedule change control system
- .3 Performance measurement
- .4 Project management software
- .5 Variance analysis
- .6 Schedule comparison bar charts

.3 Outputs
- .1 Schedule model data (updates)
- .2 Schedule baseline (updates)
- .3 Performance measurements
- .4 Requested changes
- .5 Recommended corrective actions
- .6 Organizational process assets (updates)
- .7 Activity list (updates)
- .8 Activity attributes (updates)
- .9 Project management plan (updates)
- ***.10 Progress curves update***

6.7 Activity Weights

.1 Inputs
- ***.1 Work breakdown structure***
- ***.2 Activity attributes***

.2 Tools and Techniques
- ***.1 Expert judgment***
- ***.2 Percentage calculation***

.3 Outputs
- ***.1 Relative weights***
- ***.2 Absolute weights***

6.8 Progress Curves Development

.1 Inputs
- ***.1 Relative weights***
- ***.2 Absolute weights***
- ***.3 Project schedule***
- ***.4 Weights distribution standard curves***

.2 Tools and Techniques
- ***.1 Mathematical analysis***
- ***.2 Project management software***

.3 Outputs
- ***.1 Progress curves***
- ***.2 Progress curves management plan***

6.9 Progress Monitoring

.1 Inputs
- ***.1 Progress curves***
- ***.2 Work results***
- ***.3 Schedule monitoring***

.2 Tools and Techniques
- ***.1 Progress curves development tools and techniques***
- ***.2 Project measurement criteria***
- ***.3 Progress curve analysis***

.3 Outputs
- ***.1 Actual progress***
- ***.2 Progress monitoring curves***
- ***.3 Schedule updates***
- ***.4 Corrective action***
- ***.5 Lessons learned***

Figure 6-1. Project Time Management Overview

.7 **Constraints**

In construction projects, contractual milestones are common. They involve specific events which are initially fixed and are considered as constraints.

6.1.2 Activity Definition: Tools and Techniques

See Section 6.1.2 of the *PMBOK® Guide*—Third Edition, with the additional construction-related information for Section 6.1.2.1 and the addition of Section 6.1.2.6 as follows:

.1 **Decomposition**

Care should be given not to break down the project elements into too many components, in order to prevent the creation of an unmanageable level of detail. However, it is also important that this breakdown is to a component level sufficient to ensure the work can be managed effectively. The level of detail should be appropriate to the needs of the particular project.

.6 **Concurrent Engineering**

This technique is based on constructability analysis and is commonly used on Engineering-Procurement-Construction (EPC) projects. It involves the mobilization of construction experts in the design phase to help engineering specialists decide design options such as materials, construction techniques, and plant layout arrangements. The goal with this technique is to have the least expensive, fastest, and easiest-to-build design possible, which still meets all of the functional requirements.

6.1.3 Activity Definition: Outputs

See Section 6.1.3 of the *PMBOK® Guide*—Third Edition.

6.2 Activity Sequencing

See Section 6.2 of the *PMBOK® Guide*—Third Edition.

6.2.1 Activity Sequencing: Inputs

See Section 6.2.1 of the *PMBOK® Guide*—Third Edition.

6.2.2 Activity Sequencing: Tools and Techniques

See Section 6.2.2 of the *PMBOK® Guide*—Third Edition, with the addition of Section 6.2.2.6 as follows:

.6 **Linear Scheduling Method (LSM)**

This method of scheduling is often utilized on linear projects such as highways, pipelines, etc. where station or mile-post numbers depict the physical location of the work. Linear scheduling allows activities to be visually represented at a physical location.

6.2.3 Activity Sequencing: Outputs

See Section 6.2.3 of the *PMBOK® Guide*—Third Edition.

6.3 Activity Resource Estimating

See Section 6.3 of the *PMBOK® Guide*—Third Edition.

6.3.1 Activity Resource Estimating: Inputs

See Section 6.3.1 of the *PMBOK® Guide*—Third Edition.

6.3.2 Activity Resource Estimating: Tools and Techniques

See Section 6.3.2 of the *PMBOK® Guide*—Third Edition.

6.3.3 Activity Resource Estimating: Outputs

See Section 6.3.3 of the *PMBOK® Guide*—Third Edition.

6.4 Activity Duration Estimating

See Section 6.4 of the *PMBOK® Guide*—Third Edition.

6.4.1 Activity Duration Estimating: Inputs

See Section 6.4.1 of the *PMBOK® Guide*—Third Edition.

6.4.2 Activity Duration Estimating: Tools and Techniques

See Section 6.4.2 of the *PMBOK® Guide*—Third Edition.

6.4.3 Activity Duration Estimating: Outputs

See Section 6.4.3 of the *PMBOK® Guide*—Third Edition.

6.5 Schedule Development

See Section 6.5 of the *PMBOK® Guide*—Third Edition.

6.5.1 Schedule Development: Inputs

See Section 6.5.1 of the *PMBOK® Guide*—Third Edition, with the addition of Section 6.5.1.10 as follows:

.10 Constraints

- **Imposed Dates, Major Milestones, and Key Events.** The project completion date, which is an imposed date on most construction contracts, is a significant constraint. This is the date when all contract scope should be turned over to the client and is usually referred to as the completion date, commercial operation date, provisional acceptance or similar term. A post-project completion date is often defined in terms of a period after substantial completion, and is called the final acceptance date or contract completion. This is the point in time when the contract is actually closed and only equipment and as-constructed facility warranty aspects may remain.

- **Statutory Requirements.** Development of the schedule needs to consider any limits and restrictions or other obligations placed on the project by municipal, regional, national, or international regulations. For example, there may be load limits on road travel during certain times of the year or migratory habit restrictions. Further, the incorporation of potential risk events should also be built in as possible time contingencies.

6.5.2 Schedule Development: Tools and Techniques

See Section 6.5.2 of the *PMBOK® Guide*—Third Edition, with the addition of Section 6.5.2.11 as follows:

.11 Linear Scheduling Method

This method of scheduling is utilized on linear projects such as highways, pipelines, etc. Linear scheduling allows activities to be visually represented at a physical location.

6.5.3 Schedule Development: Outputs

See Section 6.5.3 of the *PMBOK® Guide*—Third Edition.

6.6 Schedule Control

See Section 6.6 of the *PMBOK® Guide*—Third Edition.

6.6.1 Schedule Control: Inputs

See also Section 6.6.1 of the *PMBOK® Guide*—Third Edition, with the following additional construction-related input:

.5 Schedule Impacts

Unplanned situations occur within construction from a variety of causes and impact the schedule. There can be consequential effects to these impacts that need to be evaluated, which could require changes to the project or may result in claims issues as discussed in Chapter 16 of this extension.

6.6.2 Schedule Control: Tools and Techniques

See Section 6.6.2 of the *PMBOK® Guide*—Third Edition.

6.6.3 Schedule Control: Outputs

See also Section 6.6.3 of the *PMBOK® Guide*—Third Edition, with the addition of the following construction-related output:

.10 **Progress Curves (Updates).** A progress curve update is any modification to progress information caused by a modification in the project schedule, WBS, or both. Appropriate stakeholders must be notified as needed. Progress curves updates may or may not require adjustments to other aspects of the project plan.

6.7 Activity Weights Definition

Activity weights definition is the evaluation of activity characteristics and attributes for the purpose of assessing the contribution of each particular project activity to the overall project progress or to the progress of a given phase or deliverable of the project.

6.7.1 Activity Weights Definition: Inputs

.1 **Work Breakdown Structure (WBS) and Dictionary**

The WBS is the primary input to activity definition (see Section 5.3 of the *PMBOK® Guide*—Third Edition for a more detailed discussion of the WBS).

.2 **Activity Attributes**

In the context of the activity weights definition, activity attributes are those characteristics which are common for a group of activities. Durations, costs, labor hours, and quantities are examples of activity attributes. Other attributes include:

- **Duration Estimates and the Basis of Estimates.** These estimates show which characteristics of an activity drives its duration. That aspect should be taken into account when determining the activity weights (see Section 6.4 of the *PMBOK® Guide*—Third Edition for a more detailed discussion of activity duration estimating).
- **Resource Requirements and Rates.** Resource requirements and rates for each activity may be used in conjunction with duration estimates for determining the activity weights (see Chapter 7 on Project Cost Management of the *PMBOK® Guide*—Third Edition for a more detailed discussion of the resource requirements and rates).

6.7.2 Activity Weights Definition: Tools and Techniques

.1 **Expert Judgment**

Specialists can determine which activity attribute should be used for determining activity weights for each level of the WBS. In the first levels of the WBS, the attribute is usually the deliverable cost. When the decomposition level is sufficient to identify another attribute that is common to all activities in that level, the attribute should be used. When the decomposition level reaches project activities, there is usually more than one common attribute and expert judgment is needed to determine which attribute to use.

.2 **Percentage Calculation**

Based on the attributes used to determine the weight of each project deliverable or activity, a summation of those attributes is made and transformed into a percentage.

6.7.3 Activity Weights Definition: Outputs

.1 **Relative Weights**
Relative weights are percentage weights to the activities that are decomposed from a project. The relative weight of an activity represents its contribution to the project.

.2 **Absolute Weights**
Absolute weights are weights that are calculated by multiplying the relative weight of each deliverable to the total weight of the project. The absolute weight of an activity represents its contribution to the production of the overall project.

6.8 Progress Curves Development

Progress curves development is the creation of a progress baseline. This is accomplished in a similar manner to the way a cost baseline is created. See Chapter 7 on Project Cost Management (in the *PMBOK® Guide*—Third Edition and this extension) for a more detailed discussion of cost baseline. Actual progress is then plotted against the baseline as the project progresses providing an ongoing trend line which can be very helpful in forecasting future progress.

6.8.1 Progress Curves Development: Inputs

.1 **Relative Weights**
Relative weights are discussed in Section 6.7.3.2 of this extension.

.2 **Absolute Weights**
Absolute weights are discussed in Section 6.7.3 of this extension.

.3 **Project Schedule**
The project schedule determines the start and finish dates for each project activity, and hence distributes its weights in time. See Section 6.5 on Schedule Development (in the *PMBOK® Guide*—Third Edition and this extension) for a more detailed discussion about project schedule.

.4 **Weights Distribution Standard Curves**
Each activity weight is calculated based on a particular activity attribute, such as man-hours consumption, material, or cost applied. For example, the length of time for back filling an area is a function of the volume of soil deposited in that area; the soil deposition rate is determined by the equipment capabilities and is linearly distributed along the activity duration.

6.8.2 Progress Curves Development: Tools and Techniques

.1 **Mathematical Analysis**
Mathematical analysis is used to calculate the weight distribution along the project duration. Each activity has its weight and it is utilized according to standard curves. Computing the weight completed for each activity in a work

period gives the overall project progress for that work period. Repeating the analysis for all project work periods gives the overall project progress curve.

.2 **Project Management Software**

Project management software can be used to automate the process of performing the mathematical analysis. See Section 6.5.2.7 on Project Management Software (in the *PMBOK® Guide*—Third Edition and this extension) for a more detailed discussion.

6.8.3 Progress Curves Development: Outputs

.1 **Progress Curves**

Progress curves are graphical representations of the progress of the project and can be represented as follows:

- **Period or Cumulative.** The reflecting period progress or cumulative progress of the project.
- **Early or Late.** The activity dates that are used in the calculations which are based on early or late dates.
- **Overall or Partial.** A progress description that is represented for the overall project or for particular WBS deliverables. In engineering-procurement-construction (EPC) projects, progress curves are usually plotted for both the overall project and the E, P, and C phases.

.2 **Progress Curves Management Plan**

A progress curve management plan describes how progress will be measured and monitored for actual progress calculations. It may also describe how changes to the progress curves will be managed, but they usually are results of schedule changes. It may be formal or informal, highly detailed or broadly framed, based on the needs of the project.

6.9 Progress Monitoring

Progress monitoring is the evaluation of the actual project progress compared to the baseline.

6.9.1 Progress Monitoring: Inputs

.1 **Progress Curves**

Progress curves are used as a basis for comparison to the baseline (see Section 6.8.3 of this extension for a more detailed discussion of progress curves).

.2 **Work Results**

The work performed at the point in time when progress information is taken.

.3 **Schedule Monitoring**

The actual start and finish dates for project activities are the basis for actual progress calculation.

6.9.2 Progress Monitoring: Tools and Techniques

.1 Progress Curves Development Tools and Techniques

Progress curves development tools and techniques are discussed in Section 6.8.2 of this extension.

.2 Progress Measurement Criteria

Progress measurement criteria are methods for determining how much progress has been accomplished for an activity provided it is under progress in the data date. Some common methods are:

- **0% to 100%.** The weight of an activity is included in progress calculations only when it is 100% completed. This applies to activities of short duration only (i.e., those of one- month duration or shorter).
- **20% to 80%.** When the activity starts, 20% of its weight can be used for progress calculations. When it finishes, the remaining 80% of its weight is used. This convention has varying percentages, such as 30% to 70% or 50% to 50%.
- **Percentage of the Activity Duration.** The percentage of the activity weight to be considered is the same as the percentage calculated by dividing the duration from the activity start to the data date by its original duration and multiplying the result by 100.
- **Percentage of Work and Remaining Duration.** Determine and input the percentage of the work complete through observation and input the remaining duration of the activity by estimation. Doing both yields accurate representation of both cost and time, or resource and time needs based on the means of loading the schedule.

Pre-established progress measurement criteria are used to avoid conflict among stakeholders when assessing project progress.

.3 Progress Curve Analysis. Progress curve analysis is the evaluation of the actual progress compared to the baseline in order to take preventive action toward accomplishing the progress for those activities. The evaluation includes examination of the activities involved and their characteristics.

6.9.3 Progress Monitoring: Outputs

Some of the following curves are used in conjunction with the earned value analysis discussed in Section 6.6.2 and Section 7.3.2 of the *PMBOK® Guide*—Third Edition.

.1 Actual Progress

The actual progress is the summation of weights, based on the pre-established progress measurement criteria, which is accomplished within a work period.

.2 Progress Monitoring Curves

The progress monitoring curves are the graphical representations of the actual progress achieved in each control period, usually compared to a progress baseline.

.3 Schedule Updates

A schedule update may be originated by an observation that progress targets are not being met on a consistent basis. The schedule update in this case will be a consequence of changes in the activity durations and basis of estimates (see

Chapter 6 of the *PMBOK® Guide*—Third Edition for a more detailed discussion on schedule updates, activity durations, and the basis of estimates).

.4 Corrective Action

Corrective action covers what is accomplished to align the expected progress performance with the project plan. Corrective action is usually taken on another aspect of the project and is then reflected in the enhanced progress accomplishment.

.5 Lessons Learned

The causes of variances, the reasoning behind the corrective action chosen, and other types of lessons learned from progress monitoring are documented, so that these form a part of the historical database for future projects of the performing organization.

Chapter 7

Project Cost Management

See the introduction to Chapter 7 of the *PMBOK® Guide*—Third Edition, which states "Project Cost Management includes the processes involved in planning, estimating, budgeting, and controlling, costs so that the project can be completed within the approved budget" and provides an overview of the following major processes:

7.1 **Cost Estimating.** Section 7.1 of the *PMBOK® Guide*—Third Edition.
7.2 **Cost Budgeting.** Section 7.2 of the *PMBOK® Guide*—Third Edition.
7.3 **Cost Control.** Section 7.3 of the *PMBOK® Guide*—Third Edition.

All of these processes are used in construction projects. Life cycle costing, together with value engineering techniques (see Section 4.3.2.5 of this extension) and constructability analysis are used in construction projects to reduce cost and time, improve quality and performance, and optimize the decision making. See Figure 7-1 for an overview of the processes and respective inputs, tools and techniques, and outputs.

7.1 Cost Estimating

See Section 7.1 of the *PMBOK® Guide*—Third Edition.

7.1.1 Cost Estimating: Inputs

See Section 7.1.1 of the *PMBOK® Guide*—Third Edition, with the additional information as it pertains to construction projects for Section 7.1.1.1 as follows:

.1 Enterprise Environmental Factors

See also Section 7.1.1.1 of the *PMBOK® Guide*—Third Edition. One environmental input specific to construction is the pre-estimating site survey. A survey of the project or work site is required to understand the site conditions, facilities available, and logistical requirements. A checklist of items can be prepared to make sure all issues are covered. These issues must be factored in the cost estimate.

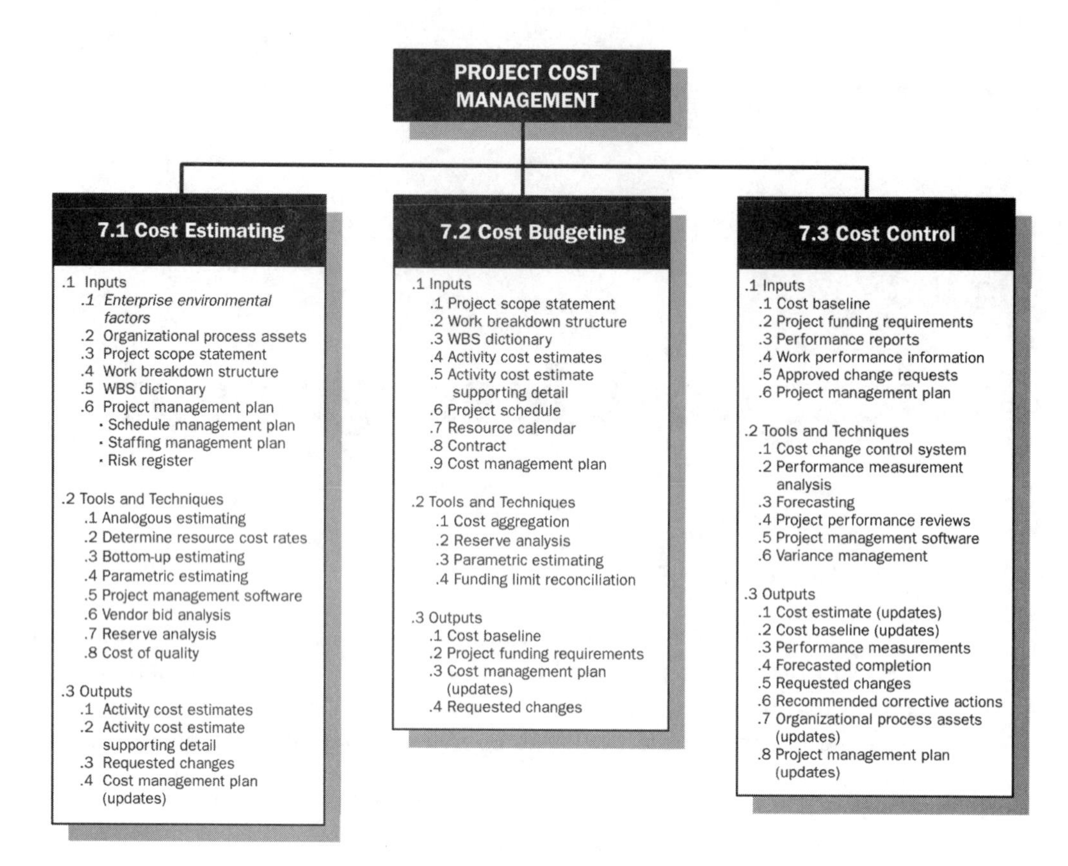

Figure 7-1. Project Cost Management Overview

7.1.2 Cost Estimating: Tools and Techniques

See Section 7.1.2 of the *PMBOK® Guide*—Third Edition.

7.1.3 Cost Estimating: Outputs

See Section 7.1.3 of the *PMBOK® Guide*—Third Edition.

7.2 Cost Budgeting

See Section 7.2 of the *PMBOK® Guide*—Third Edition.

7.2.1 Cost Budgeting: Inputs

See Section 7.2.1 of the *PMBOK® Guide*—Third Edition.

7.2.2 Cost Budgeting: Tools and Techniques

See Section 7.2.2 of the *PMBOK® Guide*—Third Edition.

7.2.3 Cost Budgeting: Outputs

See Section 7.2.3 of the *PMBOK® Guide*—Third Edition.

7.3 Cost Control

See Section 7.3 of the *PMBOK® Guide*—Third Edition.

7.3.1 Cost Control: Inputs

See Section 7.3.1 of the *PMBOK® Guide*—Third Edition.

7.3.2 Cost Control: Tools and Techniques

See Section 7.3.2 of the *PMBOK® Guide*—Third Edition.

7.3.3 Cost Control: Outputs

See Section 7.3.3 of the *PMBOK® Guide*—Third Edition.

Chapter 8

Project Quality Management

See the introduction to Chapter 8 in the *PMBOK® Guide*—Third Edition, which states, "Project Quality Management processes include all the activities of the performing organization that determine quality policies, objectives, and responsibilities so that the project will satisfy the needs for which it was undertaken." In order to satisfy this requirement in construction, an important consideration is ensuring that the project management system employs *all* processes needed to meet project requirements. Project Quality Management therefore consists primarily of ensuring that the conditions of the contract (and those contained in project technical specifications) are carried out and satisfied within the agreed-upon schedule and budget. Project Quality Management therefore addresses both the management of the project and the product of the project (and its component parts), including the assessment and determination of how the different project management processes interact to fulfill the needs of the project and whether changes or improvements are needed to accomplish the objectives of the project. Project Quality Management interacts with all other project management processes and process groups, and possibly more so than any other Knowledge Area.

Active communication with all stakeholders must be implemented to clarify what the project objectives are and the implications of their execution. On urban projects, the neighbor community is a major stakeholder, probably more so than in any other project. Special attention must be paid to the neighbor community's needs and expectations, as this can have a great impact on the project, regardless of whether or not all of the permits are obtained. Communication management is addressed in Chapter 10 of the *PMBOK® Guide*—Third Edition and this extension; therefore it will not be detailed in this chapter.

Other major stakeholders include statutory authorities, usually comprised of representative bodies from local, regional, and federal governments as is the case of regulators for the nuclear, oil, and gas industry. These regulatory bodies have their own stakeholders and respond to them in the ways described in the *Government Extension to the PMBOK® Guide*—Third Edition.

The performing organization implements the quality management system through the policy, procedures and processes of quality planning, quality assurance, and quality control, and undertakes continuous improvement activities throughout the project, as appropriate. As with safety and environmental management, quality management requires ensuring that the project management system employs all of the processes needed to meet the project requirements, and that these processes incorporate quality. Project Quality Management shares many common characteristics with Project Safety

Management and Project Environmental Management, therefore the requirements are similar. Project Quality Management, therefore, primarily ensures that the conditions of the contract (including those contained in legislation and any project technical quality specifications), are carried out to implement quality to those working on the site and to those in the vicinity of the project. Project Quality Management must therefore address the *management* of the project and the *product* of the project (and its component parts). This includes the assessment and determination of how the different project management processes interact to fulfill the needs of the project, and whether changes or improvements are needed to accomplish the quality objectives of the project. Many believe that proper and effective project management would be incomplete without due consideration of the requirements for quality management. In addition, Project Quality Management must be integrated with Risk Management processes (See Chapter 11 of this extension) to accomplish the stated objectives

Project Quality Management applies to all attributes of project management. In the construction industry, this consists of addressing three distinct (and sometimes conflicting) sets of requirements, as follows:

(a) **Mandatory statutory quality requirements imposed by legislation** and enforced by statutory third-party authorities in the region where the project is to be constructed. These are generally applicable to construction projects in specific application areas, for example, nuclear, power generation, oil, and gas industries, etc., where compliance with quality requirements are considered paramount for the safe operation of the completed facility. Similar quality requirements imposed by legislation and enforced by statutory third party authorities can also be applicable to infrastructure projects, for example, tunnels, airports, etc.

(b) **Customer quality requirements contained in the conditions of the contract,** which define how the specific quality requirements are undertaken and administered and the criteria for technical quality performance and acceptance (defined in project specifications that provide the technical quality requirements). Technical quality requirements frequently reference mandatory legislative requirements and can incorporate those for safety management (see Chapter 13) and environment management (see Chapter 14). Other requirements include those arising from enterprise environmental factors (see Section 4.1.1.3) of the *PMBOK® Guide*—Third Edition).

(c) **Specific requirements of the performing organization**, which satisfy its commercial needs (optimize profit, return on investment, increase reputation in the market place, etc.) Other requirements include those arising from organizational process assets (see Section 4.1.1.4 of the *PMBOK® Guide*—Third Edition).

The processes associated with Project Quality Management (Quality Planning, Perform Quality Assurance, and Perform Quality Control) interact with each other and with processes of other Knowledge Areas as well. Each process can involve effort from one or more persons or groups of persons based on the need and complexity of the project. Each process occurs at least once in every project and occurs in one or more project phases. The processes are presented here as discrete elements with well-defined interfaces; however, in practice they may overlap and interact in ways not detailed here. Process interactions are discussed in Chapter 3 of the *PMBOK® Guide*—Third Edition.

The requirements for the Quality Planning, Perform Quality Assurance, and Perform Quality Control activities detailed in this chapter are those generally considered applicable to construction projects most of the time. However, project sponsors or owners may invoke additional requirements, as follows:

- Constraints local to the geographical region for the project location, depending upon the scale, scope, and, complexity of the project.
- Specifications of quality management systems standards, for example, the ISO 9000 Series developed by the International Organization for Standardization (ISO), where general quality measures are considered insufficient to provide the required assurance and control.
- Industry-specific codes and standards that define specific project product performance and acceptance criteria.

It should be noted that the lack of an ISO quality management program or system does not necessarily mean the system employed by the performing organization is ineffective. Likewise, having an ISO-compliant quality management system or program does not mean the performing organization will produce a quality-compliant product.

The ISO 9000 series has been subject to significant rewriting to make it application-area specific, which is the reason for the vast array of standards associated with the ISO 9000 series. On construction projects, material testing is a common, if not a mandatory requirement. ISO has supplemented the ISO 9000 Series with ISO/IEC 17025 on General Requirements for the Competence of Testing and Calibration Laboratories, to address the unique competency and operational requirements for material testing laboratories. However, as stated previously having an ISO/IEC 17025-compliant testing laboratory management system does not equate with a project-compliant system, as customer requirements often exceed the basic requirements detailed in ISO/IEC 17025. The scope of requirements to supplement ISO 17025 to produce a project-compliant system will be very much dependent on the requirements of the industry application area, project sponsor/owner requirements, and/or the performing organization's scope of ISO 17025 accreditation.

Modern quality management complements project management. For example, both disciplines recognize the importance of the following:

.1 **Prevention Over Inspection.** See *PMBOK® Guide*—Third Edition, Chapter 8. Although verification by inspection and testing are synonymous as they pertain to construction projects, it is also important to determine the most appropriate methods of verification and verification criteria. Critical issues include but are not limited to:
   - Evaluating performance characteristics of the end product of the project (or component part) as a whole vs. inspecting or testing individual characteristics
   - Ensuring that the correct and objective data is captured, rather than arbitrary or subjective data.

.2 **Responsibility.** See *PMBOK® Guide*—Third Edition, Chapter 8. Management responsibility includes the timely provision of adequate resources, in terms of finance, human resources, and equipment, etc.

.3 **Continuous Improvement.** See *PMBOK® Guide*—Third Edition, Chapter 8. Continuous improvement is integral to project management, and needs to be integrated into the project management system as a whole.

.4 **Risk Management.** Quality, safety, and environmental management are frequently considered a subset of risk management. As such, these must address more than the technical requirements of a project. It is frequently overlooked that the contract itself is a quality, safety, and environmental management system standard. A specialized management system to address the unique characteristics of a specific construction project is frequently required, for example, the requirements for surface rail or highway infrastructure projects are fundamentally different. Such specialization would frequently require the adoption of both the requirements of this extension and ISO standards.

As stated in Chapter 8 of the *PMBOK® Guide*—Third Edition, "Quality and grade are not the same. Grade is a category assigned to products or services having the same functional use but different technical characteristics. Low quality is always a problem; low grade may not be." For example, an office building may be of high quality (e.g., having appropriate floor layout, absence of leakage, functioning mechanical, and electrical systems, etc.) and low grade (e.g., having concrete floors), or of low quality (e.g., poorly designed floor space, roof leakage, improperly balanced heating and ventilation systems, etc.) and high grade (wood trim and marble floors). The determination of and delivery of the required levels of quality and grade are the responsibilities of the project manager and the project management team. In such instances, the phrase "acceptance and performance criteria" may be a more accurate definition for the term "quality criteria." For example, where concrete is to be used in a marine or offshore environment, its ability to resist chloride penetration (to avoid rebar corrosion which could lead to catastrophic failure of a structure) could be equally as important or more important as concrete strength, as designing a concrete structure (and concrete grade) with include for factors of safety, whereas there is a degree of reliance on workmanship to ensure the concrete in placed and compacted correctly.

See Figure 8-1 for an overview of the processes and respective inputs, tools and techniques, and outputs.

8.1 Quality Planning

See Section 8.1 of the *PMBOK® Guide*—Third Edition. Quality standards are not only comprised of project codes, regulations, and standards; they also include any condition of a contract for which tangible deliverables have been defined and which will be used to determine acceptance. The contract is the principle project quality standard, as it specifies the applicable statutory and legislative quality requirements, technical quality codes, standards, and regulations.

8.1.1 Quality Planning: Inputs

See Section 8.1.1 of the *PMBOK® Guide*—Third Edition, additional inputs related to construction are included in 8.1.1.5 through 8.1.1.9 as follows:

.5 **Contract Requirements**

Contract requirements include any and all requirements specified in the contract documents for the project. Specifications, regulations, legislation, and standards (technical or legislative) are contract requirements that are specific to construction projects. The project management team must consider area-specific standards, specifications, or regulations including those arising from local, regional

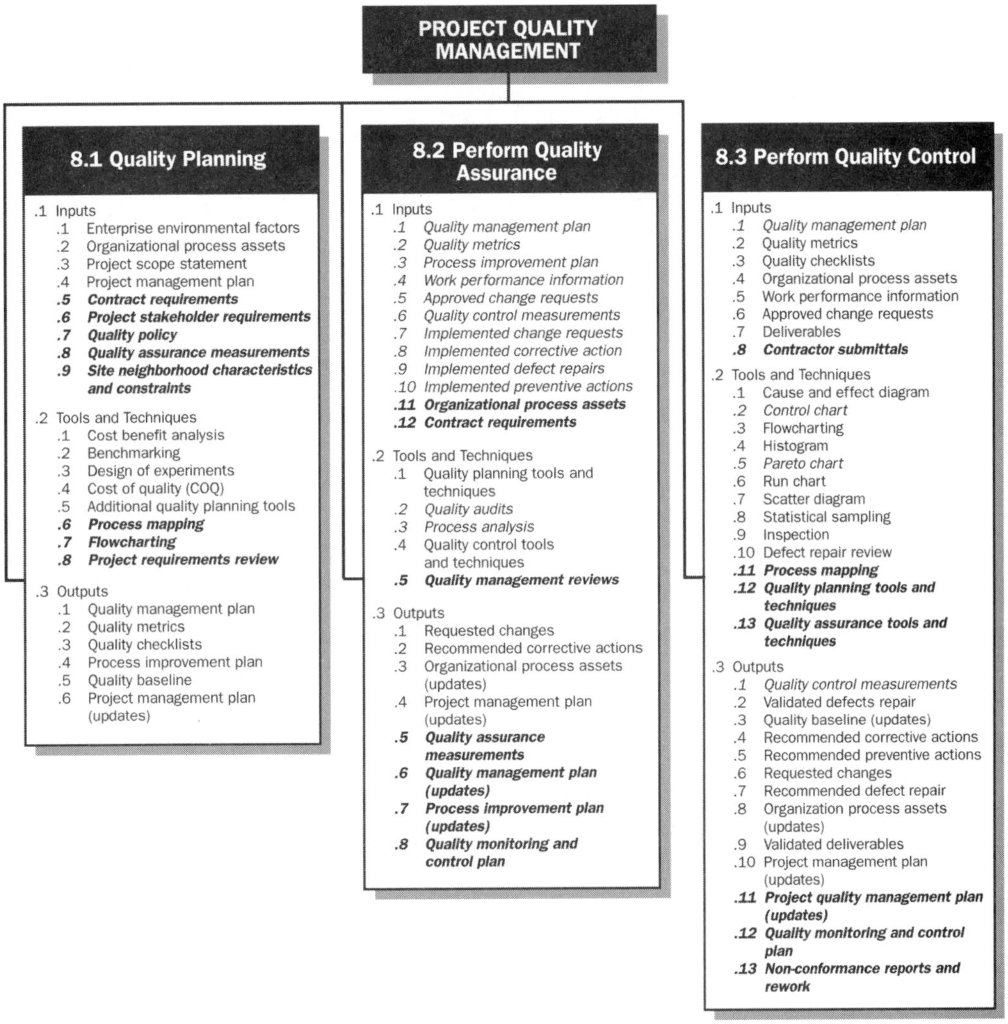

PROJECT QUALITY MANAGEMENT

8.1 Quality Planning

.1 Inputs
- .1 Enterprise environmental factors
- .2 Organizational process assets
- .3 Project scope statement
- .4 Project management plan
- ***.5 Contract requirements***
- ***.6 Project stakeholder requirements***
- ***.7 Quality policy***
- ***.8 Quality assurance measurements***
- ***.9 Site neighborhood characteristics and constraints***

.2 Tools and Techniques
- .1 Cost benefit analysis
- .2 Benchmarking
- .3 Design of experiments
- .4 Cost of quality (COQ)
- .5 Additional quality planning tools
- ***.6 Process mapping***
- ***.7 Flowcharting***
- ***.8 Project requirements review***

.3 Outputs
- .1 Quality management plan
- .2 Quality metrics
- .3 Quality checklists
- .4 Process improvement plan
- .5 Quality baseline
- .6 Project management plan (updates)

8.2 Perform Quality Assurance

.1 Inputs
- *.1 Quality management plan*
- *.2 Quality metrics*
- *.3 Process improvement plan*
- *.4 Work performance information*
- *.5 Approved change requests*
- *.6 Quality control measurements*
- *.7 Implemented change requests*
- *.8 Implemented corrective action*
- *.9 Implemented defect repairs*
- *.10 Implemented preventive actions*
- ***.11 Organizational process assets***
- ***.12 Contract requirements***

.2 Tools and Techniques
- .1 Quality planning tools and techniques
- *.2 Quality audits*
- *.3 Process analysis*
- .4 Quality control tools and techniques
- ***.5 Quality management reviews***

.3 Outputs
- .1 Requested changes
- .2 Recommended corrective actions
- .3 Organizational process assets (updates)
- .4 Project management plan (updates)
- ***.5 Quality assurance measurements***
- ***.6 Quality management plan (updates)***
- ***.7 Process improvement plan (updates)***
- ***.8 Quality monitoring and control plan***

8.3 Perform Quality Control

.1 Inputs
- *.1 Quality management plan*
- .2 Quality metrics
- .3 Quality checklists
- .4 Organizational process assets
- .5 Work performance information
- .6 Approved change requests
- .7 Deliverables
- ***.8 Contractor submittals***

.2 Tools and Techniques
- .1 Cause and effect diagram
- *.2 Control chart*
- .3 Flowcharting
- .4 Histogram
- *.5 Pareto chart*
- .6 Run chart
- .7 Scatter diagram
- .8 Statistical sampling
- .9 Inspection
- .10 Defect repair review
- ***.11 Process mapping***
- ***.12 Quality planning tools and techniques***
- ***.13 Quality assurance tools and techniques***

.3 Outputs
- *.1 Quality control measurements*
- .2 Validated defects repair
- .3 Quality baseline (updates)
- .4 Recommended corrective actions
- .5 Recommended preventive actions
- .6 Requested changes
- .7 Recommended defect repair
- .8 Organization process assets (updates)
- .9 Validated deliverables
- .10 Project management plan (updates)
- ***.11 Project quality management plan (updates)***
- ***.12 Quality monitoring and control plan***
- ***.13 Non-conformance reports and rework***

Figure 8-1. Project Quality Management Overview

and national governments, which will affect the project. Specifications, regulations, legislation and standards generally refer to either:

- Performance and acceptance criteria that pertain to the product(s) of the project.
- Workmanship criteria or how work is to be undertaken.

.6 Project Stakeholder Requirements

The requirements of project stakeholders are described in Section 2.2 of the *PMBOK® Guide*—Third Edition, with management of project stakeholder information described in Section 10.4 of the *PMBOK® Guide*—Third Edition. The quality requirements, and all selected alternatives to balance them, must be negotiated and agreed with project stakeholders, including the surrounding community and government agencies.

.7 Quality Policy

The quality management policy would also include the degree to which the performing organization's management is committed to quality, and can have a major impact on the effectiveness of a quality program.

.8 Perform Quality Assurance Measurements

Perform quality assurance measurement generates feedback on quality assurance activity performance and is fed back into the Quality Planning Process (Section 8.1.1) for use in re-evaluating and analyzing the process. This feedback may include the performance of the planning activities of the performing organization, the criteria and processes employed, and, where improvements are required. Perform quality assurance measurement input is also used in Perform Quality Control (Section 8.3.1) as an indicator of areas which may need further investigation and to re-assess the risk or decisions taken in early project phases.

.9 Site Neighborhood Characteristics and Constraints

The characteristics of the construction site and the surrounding environment must be known prior to project execution. For construction projects, the environment is the surrounding neighborhood where the project is to be undertaken, for which there may be constraints pertaining to quality management, safety management and environmental management. These can include proximity of adjacent residents, configuration of project offices, layout and location of construction equipment workshop, material delivery time constraints, traffic congestion in vicinity of the project site during peak periods, etc.

8.1.2 Quality Planning: Tools and Techniques

See Section 8.1.2 of the *PMBOK® Guide*—Third Edition, with additional construction-related tools and techniques as provided in 8.1.2.6 through 8.1.2.8.

.6 Process Mapping

Process mapping is commonly undertaken with flowcharting (see Section 8.1.2.7) to:

- Map how a particular process is carried out;
- Determine how various processes interact;
- Identify any gaps in a particular work item or activity (termed gap analysis), and include the absence of critical review points or a required deliverable (including the omission of verification that work has been undertaken and is acceptable. Such verification could relate not only to quality, but also safety or environmental issues).

.7 Flowcharting

A flowchart is a diagram that shows how various elements of a system or process (or series of processes) interact or interrelate. Flowcharting is commonly used with process mapping (see Section 8.1.2.6) and with certain process statistical analyses and reporting methods. Flowcharting is also used as a means of identifying non-value-added activities or functions, or finding delay points or bottlenecks in tasks or activities.

.8 Project Requirements Review

Project requirements review includes an assessment and determination of the following:

- The characteristics and criteria of each component part of the product(s) of the project, and how to satisfy them;
- The applicable verification criteria, including those required to demonstrate that acceptance and performance characteristics are fulfilled;

- Alternatives Review and Selection. In construction projects it is not uncommon for some activities to be performed with different processes or arrangements for achieving the same result or output. This applies equally to quality management. Examples include:
  - Rock formations can be removed by blasting or by using pneumatic breakers;
  - Effluents from chemical pipe cleaning can be treated at an on-site waste treatment works, or taken to an external treatment facility;
  - Materials can be tested at an on-site laboratory, or sent to an external laboratory for testing;
  - Inspections can be carried out by independent inspection organizations or undertaken by those carrying out the work (provided the competence of the latter to do so has been determined, assessed and agreed).

Another increasingly common example is where a requirement (standard or specification) developed in one geographical region is to be employed in another result in characteristics not common in the region of use. This is where trade-offs are frequently necessary, and the re-qualification of the requirements is necessary to meet the quality objectives. The term "trade-off" does not imply lowering standards for quality, but to show the same end result can be achieved in different ways. Trade-offs need to be carefully scrutinized so as not to compromise quality, or other requirement, and are rarely accepted without valid justification by project sponsors or owners, for obvious reasons.

Generally, all processes are analyzed to determine alternatives to increase effectiveness and efficiency, for example, cost-benefit analyses (see Sections 8.1.2.4 and 8.3.2.1) and others where time, cost, quality, safety and environmental aspects need to be balanced or even exceeded. Quality requirements can be mandatory constraints, as non-compliances can cause the project to have its permits cancelled or revoked.

While most project stakeholders will participate during the elaboration process, some (usually government bodies) may require the information to be formally issued before an analysis is made. In such cases, conformance to laws, regulations, standards, and, specifications, should be sufficient to obtain the requisite approvals.

8.1.3 Quality Planning: Outputs

See Section 8.1.3 of the *PMBOK® Guide*—Third Edition.

8.2 Perform Quality Assurance

See Section 8.2 of the *PMBOK® Guide*—Third Edition. Perform Quality Assurance involves the following:

- Applying the planned, systematic quality activities to ensure that the project employs all processes needed to meet quality requirements;
- Determining whether these processes (and their integration) are effective in ensuring the project management system will fulfill the quality requirements of the project and the product of the project; and
- Evaluating the results of quality management on a regular basis to provide confidence that the project will satisfy the relevant quality standards.

8.2.1 Perform Quality Assurance: Inputs

See Section 8.2.1 of the *PMBOK® Guide*—Third Edition, with the modification of 8.2.1.1 through 8.2.1.10 and the additions of 8.2.1.11 and 8.2.1.12:

.1 Quality Management Plan

See Section 8.1.3.1 of the *PMBOK® Guide*—Third Edition and 8.1.3.1 of this extension. The quality management plan describes how quality assurance will be applied and performed on the project and defines the primary attributes of the project quality management plan which form inputs to the Perform Quality Assurance process.

.2 Quality Metrics

See Section 8.1.3.2 of the *PMBOK® Guide*—Third Edition.

.3 Process Improvement Plan

See Section 8.1.3.4 of the *PMBOK® Guide*—Third Edition.

.4 Work Performance Information

See Sections 4.4.3.7 and 10.3.3.1 of the *PMBOK® Guide*—Third Edition.

.5 Approved Change Requests

See Sections 4.4.1.4 and 8.2.1.5 of the *PMBOK® Guide*—Third Edition.

.6 Quality Control Measurements

See Sections 8.2.1.6 and 8.3.3.1 of the *PMBOK® Guide*—Third Edition.

.7 Implemented Change Requests

See Section 4.4.3.3 of the *PMBOK® Guide*—Third Edition.

.8 Implemented Corrective Actions

See Section 4.4.3.4 of the *PMBOK® Guide*—Third Edition.

.9 Implemented Defect Repair

See Section 4.4.3.6 of the *PMBOK® Guide*—Third Edition.

.10 Implemented Preventive Actions

See Section 4.4.3.5 of the *PMBOK® Guide*—Third Edition.

.11 Organizational Process Assets

See Section 4.1.1.4 of the *PMBOK® Guide*—Third Edition.

.12 Contract Requirements

See Section 8.1.1.5 of the *PMBOK® Guide*—Third Edition.

8.2.2 Perform Quality Assurance: Tools and Techniques

See Section 8.2.2 of the *PMBOK® Guide*—Third Edition, with the modification of 8.2.2.2 and 8.2.2.3; and the addition of 8.2.2.5.

.2 Quality Audits

See also Section 8.2.2.2 of the *PMBOK® Guide*—Third Edition. Audits involve structured and independent reviews to demonstrate whether project activities of the performing organization(s) comply with the project requirements and whether such activities are suitable to fulfill the requirements of the project.

Audits of the project product(s) and/or its component parts are sometimes termed "quality technical or quality compliance audits" and include an evaluation of results or outputs of work activities compared to the performance and acceptance criteria defined in technical quality standards and specifications.

Quality audits can also be performed for the project management system as a whole or of its individual component parts, that is, the procurement management system (refer to Section 12.5.2.3 of the *PMBOK® Guide*—Third Edition), design management system, or commissioning management system, etc.

Audits are also carried out where compliance with management systems standards is required, such as ISO 9001. ISO has developed the ISO 10011 series for quality auditing.

Integrated audits are commonly adopted (for example incorporating the applicable requirements such as those for quality, safety and environmental management) to afford a more accurate measure of the effectiveness of a specific area of work in fulfilling project requirements. and are undertaken to assess the effectiveness of the controls employed on a project as a whole rather than individually.

.3 Process Analysis

See Section 8.2.2.3 of the *PMBOK® Guide*—Third Edition. Process analyses may also include statistical analyses (see Section 8.3.2.1 of the *PMBOK® Guide*—Third Edition).

.5 Quality Management Reviews

Quality management reviews provide an assessment and evaluation of the effectiveness and suitability of the project management system as whole or in part by the performing organization's top management. Results of quality management reviews are used to effect changes and improvements to those elements of the project management system that are not performing satisfactorily.

8.2.3 Perform Quality Assurance: Outputs

See Section 8.2.3 of the *PMBOK® Guide*—Third Edition, with the addition of Sections 8.2.2.5 through 8.2.2.8:

.5 Quality Assurance Measurements

Perform quality assurance measurements is the result of perform quality assurance activities that are fed back into the Quality Planning process see Section 8.1) for use in re-evaluating and analyzing the performance of the performing organization, and the standards and processes employed. Perform Quality Assurance measurements are also used as input into Perform Quality Control (see Section 8.3) as an indicator of areas, which may need further investigation.

.6 Quality Management Plan (Updates)

See Section 8.1.3.1 of the *PMBOK® Guide*—Third Edition.

.7 Process Improvement Plan (Updates)

See 8.1.3.4 of the *PMBOK® Guide*—Third Edition.

.8 Quality Monitoring and Control Plan

The quality monitoring and control plan describes how the project management team will implement the necessary quality controlling activities of the performing organization. The quality monitoring and control plan is a component or subsidiary plan of the project management plan (see Section 4.3 of the *PMBOK® Guide*—Third Edition). The quality monitoring and control plan will either contain or make reference to specific procedures that are to be employed for ensuring the quality compliance of the work performed.

While the project quality management plan details how the performing organization will perform management quality on the project, the project quality monitoring and control plan defines the actual monitoring and control activities to be employed, especially the following:

- Item of work to be monitored;
- Reference to the applicable reference document and acceptance criteria;
- Applicable verification activities that will be performed, and when such activities are performed in relation to the overall process;
- Responsibility for the work and each verification activity;
- Applicable characteristics and measurements which are taken or recorded;
- Applicable supporting documentation which is generated to demonstrate satisfactory or unsatisfactory performance.

To be effective, quality monitoring and control (and verification) must be integrated into how the physical work will be carried out. This process establishes control points or gates throughout the process to ensure the next phase work cannot proceed till the preceding work has been completed and verified as complete and compliant.

An emerging trend worldwide is the use of integrated execution and verification plans (EVPs) that incorporate (integrate) the necessary verification activities into actual work processes. EVPs also serve as plan work processes and assist in determining the sequence by which such work needs to be undertaken. A similar approach can be adopted for monitoring and controlling activities for Safety Management (see Section 13 of this extension) and Environmental Management (see Section 14 of this extension), as EVP contains the precise sequence of work to be undertaken and all necessary monitoring and controlling verification activities.

8.3 Perform Quality Control

See Section 8.3 of the *PMBOK® Guide*—Third Edition. Perform quality control involves the following:

- Determining and applying measures for monitoring achievement of specific project results throughout the project to determine whether they comply with the requirements, identifying unsatisfactory performance;
- Identifying ways to eliminate causes of unsatisfactory performance. This includes identifying failures on the part of quality planning and quality assurance.

8.3.1 Perform Quality Control: Inputs

See Section 8.3.1 of the *PMBOK® Guide*—Third Edition, with the modification of 8.3.1.1 and the addition of 8.3.1.8, as follows:

.1 Quality Management Plan.
See also Section 8.1.3.1 of the *PMBOK® Guide*—Third Edition and Section 8.2.1.1 of this extension.

.8 Contractor Submittals.
Refer to Section 5.5.1 of this extension.

8.3.2 Perform Quality Control: Tools and Techniques

See Section 8.3.2 of the *PMBOK® Guide*—Third Edition, with the modification of Sections 8.3.2.2 and 8.3.2.5 and the addition of 8.3.2.11 through 8.3.2.13, as follows:

.2 Control Charts
See Section 8.3.2.2 of the *PMBOK® Guide*—Third Edition, which provides a graphic representation of the mean value of a measurement against the Upper and Lower Control limits (UCL and LCL). Control charts also define the Upper and Lower Action Limits (UAL and LAL), which are used as triggers or indicators to forewarn the performing organization of a process is approaching unsatisfactory performance.

.5 Pareto Chart
See Section 8.3.2.5 of the *PMBOK® Guide*—Third Edition. Pareto analyses are commonly used to assess/evaluate the finding of quality audits (see 8.2.2.2) to show where the root cause(s) of unsatisfactory performances originate.

.11 Process Mapping
See Section 8.1.2.6 of this extension.

.12 Quality Planning Tools and Techniques
See Section 8.1.2 of the *PMBOK® Guide*—Third Edition.

.13 Quality Assurance Tools and Techniques
See Section 8.2.2 of the *PMBOK® Guide*—Third Edition.

8.3.3 Perform Quality Control: Outputs

See 8.3.3 of the *PMBOK® Guide*—Third Edition, with modifications as described in 8.3.3.1 and additional outputs as described in 8.3.3.11 through 8.3.3.13, as follows:

.1 Quality Control Measurements
See Section 8.3.3.1 of the *PMBOK® Guide*—Third Edition, which represents the results of quality control activities that are fed back as inputs into the Perform Quality Planning (see Section 8.1.1) and Perform Quality Assurance (See Section 8.2.1) processes of the performing organization to re-evaluate and analyze the performance of the project quality management system, including applicable quality standards in contract requirements (see Section 8.1.1.5).

.11 Project Quality Management Plan (Updates)

See Section 8.1.3.1 of the *PMBOK® Guide*—Third Edition, including updates to those attributes described in Section 8.2.1.1 of this extension.

.12 Quality Monitoring and Control Plan

The quality monitoring and control plan describes how the project management team will implement the necessary quality controlling activities of the performing organization. The quality monitoring and control plan is a component or subsidiary plan of the project management plan (see Section 4.3 of the *PMBOK® Guide*—Third Edition). The quality monitoring and control plan will either contain or make reference to specific procedures that are to be employed for ensuring the quality compliance of the work performed.

While the project quality management plan details how the performing organization will perform quality management on the project, the project quality monitoring and control plan defines the actual monitoring and control activities to be employed, especially the following:

- Item of work to be monitored;
- Reference to the applicable reference document and acceptance criteria;
- Applicable verification activities that will be performed, and when such activities are performed in relation to the overall process;
- Responsibility for the work and each verification activity;
- Applicable characteristics and measurements which are taken or recorded;
- Applicable supporting documentation which is generated to demonstrate satisfactory or unsatisfactory performance.

To be effective, quality monitoring and control (and verification) must be integrated into how the physical work will be carried out. This process establishes control points or gates throughout the process to ensure the next phase work cannot proceed till the preceding work has been completed and verified as complete and compliant.

An emerging trend worldwide is the use of integrated execution and verification plans (EVPs) that incorporate (integrate) the necessary verification activities into actual work processes. EVPs also serve as plan work processes and determine the sequence by which such work needs to be undertaken. A similar approach can be adopted for monitoring and controlling activities for safety management (see Section 13 of this extension) and environmental management (see Section 14 of this extension), whereas EVP contains the precise sequence of work to be undertaken and all necessary monitoring and controlling-verification activities.

.13 Nonconformance Reports and Rework

Items that are inspected and found not to be in compliance with requirements commonly have a non-conformance report prepared outlining the deficiencies, the immediate corrective action to bring the non-conforming work within permissible limits of tolerances, and the preventive action to prevent recurrence of the condition that caused the non-conformance. Non-conformance reports take many forms, for example:

- *Field Deficiency Reports* (FDRs) used to record product or workmanship defects. Repeat field deficiencies could lead to the matter being elevated to the status of a Non-Conformance Report, as this would indicate problems with the process or system being employed;

- *Non-conformance Reports* (NCRs), used to record system deficiencies, such as those identified during audits Repetitive non-conformance issues could result in the matter being elevated to that of a contract violation notice, because this would indicate the existence of problems with the manner in which component parts of the project's product are being managed;
- *Contract Violation Notices* (CVNs), used to record principle contract document violations, which would indicate failure on the part of the performing organization regarding the overall requirements of the contract.

Rework is the action taken to bring a defective or non-conforming item back into compliance with requirements or specifications. Rework, especially unanticipated rework, is a frequent cause of cost and schedule overruns on construction projects.

The project team should make every reasonable effort to minimize rework. Often the cause of rework can be prevented or minimized by the imposition of an approved quality control program at a supplier's plant. These supplier quality programs are often developed jointly between the performing organization and the supplier, and can also have involvement from the project sponsor/owner, other applicable project stakeholders, and any applicable industry regulator.

Chapter 9

Project Human Resource Management

See the introduction to Chapter 9 in the *PMBOK® Guide*—Third Edition, which states "Project Human Resource Management includes the processes required to organize and manage the project team. The project team is comprised of the people who have assigned roles and responsibilities for completing the project." See Figure 9-1 for an overview of the processes and respective inputs, tools and techniques, and outputs, which includes a fifth construction-related process, Close Project Team, because of the importance of project closing functions to resource management in construction projects. The project workforce, as it relates to construction projects, can be divided into managerial and labor forces. The labor force (construction trades), is the largest component of the site workforce, generally much larger than the managerial force. This section covers both workforce components.

One of the distinguishing features of human resource management in a construction project is the fact that the project location is almost always unique to the project and is not "home." The project team is often not working in the familiar environment of their home office, but rather in the artificial environment of a construction site. To some degree, allegiance is necessarily transferred from "home" to the project site, its temporary management, rules, culture, geography, etc. This situation creates additional and important demands on the function of human resource management beyond those described in the *PMBOK® Guide*—Third Edition and can require unique methods of dealing with them. As discussed later in this section, the methods and procedures for acquiring labor for a construction project can vary significantly in different parts of the world, and managers of construction projects need to be aware of local conditions and customs.

9.1 Human Resource Planning

See Section 9.1 of the *PMBOK® Guide*—Third Edition, which states, "Human resource planning determines project roles, responsibilities and reporting relationships, and creates the staffing management plan. Project roles can be designated for persons or groups. Those persons or groups can be from inside or outside of the organization performing the project." In construction, human resource planning includes the sub-

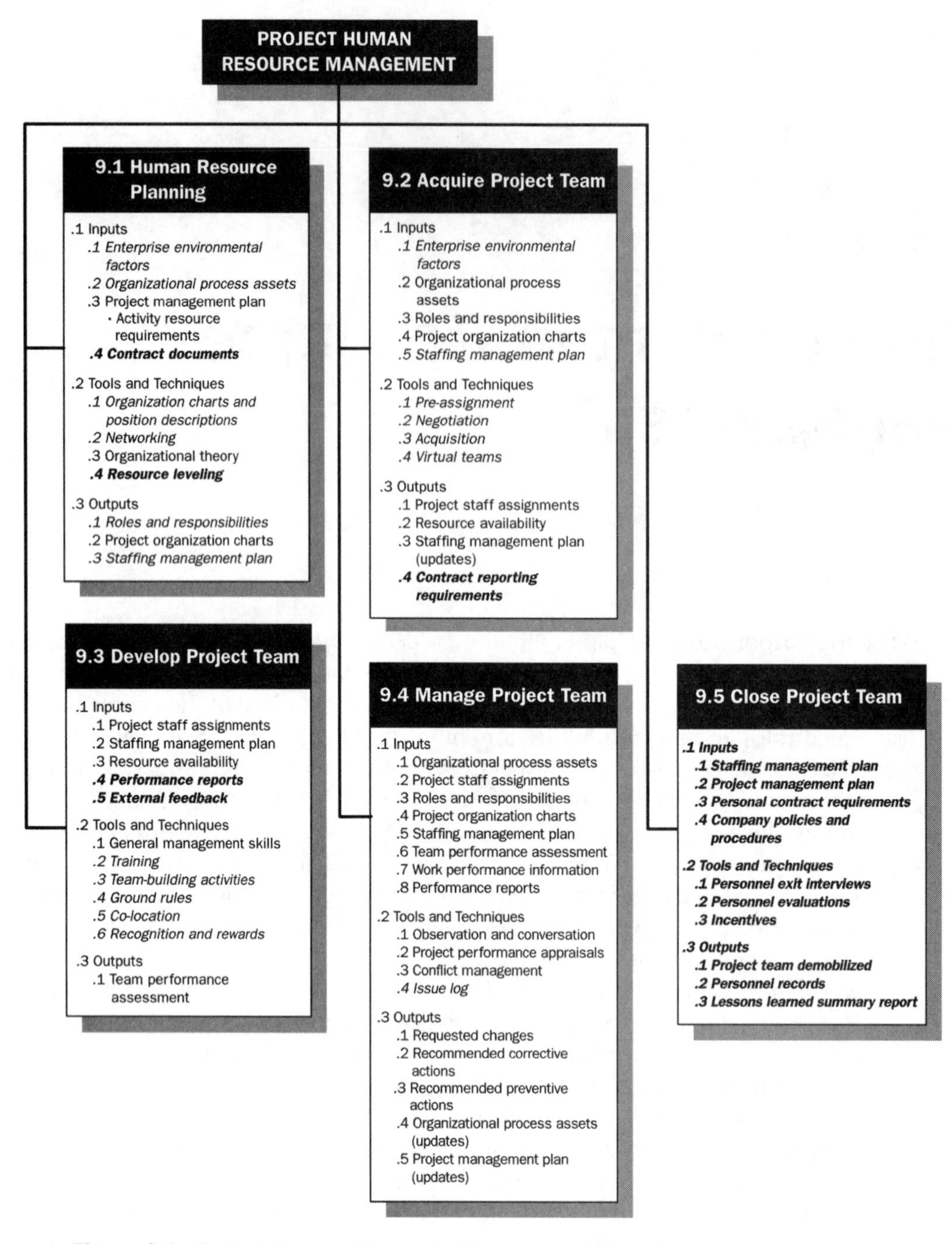

Figure 9-1. Project Human Resource Management Overview

contractors for construction work or services; it also includes the assessment of general requirements and availability of the labor workforce.

9.1.1 Human Resource Planning: Inputs

See 9.1.1 of the *PMBOK® Guide*—Third Edition, with the modification of 9.1.1.1 and 9.1.1.2 and in the addition of Section 9.1.1.4 as follows:

.1 Enterprise Environmental Factors

See Section 9.1.1.1 and Section 4.1.1.3 in the *PMBOK® Guide*—Third Edition. In construction, some of the relevant enterprise environmental factors (Section 4.1.1.3 of this extension) involving organizational culture and structure are:

- **Organizational interfaces.** Among several other factors, these interfaces include formal and informal reporting relationships or interfaces. The organizational interface for most design and construction projects is usually well defined as a result of long experience with this process: owner, designer, contractor, subcontractors, finance and regulatory agencies, and the general public. The primary interface is usually between the owner, designer, and contractor(s). Organizational interfaces can become critical in effectively managing certain contract forms such as design-build and construction management. The project stakeholders' roles and responsibilities must be clearly identified at the project outset. In addition, variations in contract agreements between stakeholders, such as design-build and contractor joint ventures can add other complexities to the stakeholder internal interfaces.
- **Project team preferences.** These include constraints that limit the project team's options, including historical preferences among members of the project team. In construction, owners often have preferences for union or non-union construction forces, certain suppliers, and contractors as a result of trade relations.
- **Funding-related constraints.** Many publicly funded projects have tight workforce and subcontractor requirements for Minority/Woman/Emerging Small Business (M/W/ESB) or disadvantaged business enterprise standards, jobsite training, and apprenticeship programs, as well as stipulations for ethnicity and minorities. Similarly, local governments may dictate that certain companies or individuals should be employed in various roles and capacities for public projects. Projects funded by international aid or financing organizations may have specific requirements relating to the use of local labor. All these constraints can affect project alternatives and should be accounted for in the Project Management Plan (Section 4.3.3.1 of the *PMBOK® Guide*, Third Edition), as well as in the management of project human resources.
- The owner (or client) often approves portions of the organizational plan either in the proposal or contract and it may be difficult to change the structure later without the owner's approval.
- **Technical.** In addition to the engineering and construction technical requirements, many projects require specialized considerations for the logistics and safety of both the labor force and members of the general public who may come in contact with the project. This includes the health and sometimes welfare of the employees.
- **Interpersonal.**
- **Logistical.**
- **Political.**

.2 Organizational Process Assets

See Section 9.1.1.2 and 4.1.1.3 of the *PMBOK® Guide*—Third Edition. Many organizations have worked diligently to acquire and assemble the talent and experience represented on a lead project team and do their best to keep these resources as employees. Coping with and maintaining the experienced staff during the rise and fall of project resource needs is a continual concern for these organizations.

In addition to various templates and checklists, many organizations rely upon key individuals with significant personal experience to forecast the project resources needed for technical and logistical purposes. The project schedule often becomes one of the key planning tools for the project site superintendent to forecast labor requirements. See also Chapter 6 on Time Management and Chapter 12 on Project Procurement Management.

.4 Contract Documents

The construction industry utilizes project-specific contracts that comprise a part of the contract documents, providing project requirements pertaining to the prevailing wage rates and other project labor requirements, such as certified payrolls or federal or international standards. Most often these requirements are part of the general conditions and/or supplementary conditions, which call out the contractor's responsibilities, including applicable laws and regulations, permits, site conditions, safety, hazards communication programs, and other requirements pertaining to subcontractors, suppliers, and others. See also Section 4.1.1.4 which covers constraints surrounding the project plan development, which are often spelled out in the contract documents, and Section 12.1 regarding procurement planning.

9.1.2 Human Resource Planning: Tools and Techniques

See Section 9.1.2 of The *PMBOK® Guide*—Third Edition, with the modification of Sections 9.1.2.1 and 9.1.2.2 and the addition of Sections 9.1.2.4.

.1 Organizational Charts and Position Descriptions

See Section 9.1.2.1 of the *PMBOK® Guide*—Third Edition and see also Claim Prevention (Section 16.3 of this extension). Organizational charts and position descriptions are particularly important on a construction project, where most of the team is located, to provide a clear understanding of each member's role and competency, ensuring a smoother interpersonal interface. In addition, it is beneficial for the project team to understand the site level of authority for each organization involved on the project. Project partnering as a technique for dispute avoidance can also be a great benefit to the project team and organizations that must deal with the many potential areas of the contract documents in which interpretation of a requirement can often lead to a dispute. See Section 16.3.1 of this extension.

.2 Networking

See also Section 9.1.2.2 of the *PMBOK® Guide*—Third Edition. Networking within the construction industry is very important in terms of learning who the stakeholders are in certain markets, understanding the capacity of subcontractors and identifying potential vendors, suppliers, designers, engineers, and other subcontractors who may bid on projects.

.4 Resource Leveling

See also Section 6.3.1 of the *PMBOK Guide®*—Third Edition. Resource leveling is often used to keep the workforce as productive as possible because construction projects may have a large labor workforce and the financial success of the project can be closely linked to productivity. Resource leveling assigns the necessary workforce to each line item on the project schedule, projecting the total labor force employed on a period basis. The output from this activity may indicate an unacceptable variation in the total force or in some of the trades components. In most cases, the resources cannot be hired and fired on a regular basis without negatively affecting the project's productivity and costs, due to excessive re-orienting and training. Non-critical path items should be adjusted to level the manpower required to the extent possible. Activity resource estimating within the project scheduling process must consider the availability of the resources and any contract requirements and/or site conditions that could have an effect on resource capacity.

9.1.3 Human Resource Planning: Outputs

See Section 9.1.3 of the *PMBOK® Guide*—Third Edition, with modifications to Sections 9.1.3.1 and 9.1.3.3 as follows:

.1 Roles and Responsibilities

See Section also 9.1.3.1 of the *PMBOK® Guide*—Third Edition. Project roles and responsibilities must be assigned to the appropriate project stakeholders. Roles and responsibilities may vary over time. Most roles and responsibilities will be assigned to stakeholders who are actively involved in the work of the project, such as the project/construction manager, project manager, lead architect, superintendent or inspectors, and other members of the project management team, and the individual contributors. This is usually done informally on small projects. However, it becomes a formalized organizational document on large projects, especially multi-national project teams. Project roles and responsibilities should be closely linked to the project scope definition and the Work Breakdown Structure (WBS). A Responsibility Assignment Matrix (RAM) can be prepared for this purpose. See Figure 9-5 in the *PMBOK® Guide*—Third Edition.

.3 Staffing Management Plan

See also Section 9.1.3.3 of the *PMBOK® Guide*—Third Edition. In addition to the production aspects of the plan, serious considerations must be incorporated regarding the project quality, safety, health, and environmental staffing that support the construction operations and guarantee its performance. While there may be limited requirements where the construction project is in an urban area serviced by the proper welfare services in developed countries, the same would not be present in many developing countries or indeed in rural or remote settings. Requirements to be considered are provisions, maintenance of work infrastructure, work-crew camps, canteens, transportation, associated utilities, security, and medical facilities.

The staffing management plan is a working component, much like the project schedule, and needs to incorporate the changing project needs. Most plans consist of permanent company team members and additional personnel are hired locally to fill needed positions. Unexpected events can have a great impact on construction projects when they occur after the staffing plan was initially created. For example, a team member may leave the project for various reasons, and a vital position would be void. In cases where the project is in unfamiliar territory, the replacement of this resource can be an expensive process.

9.2 Acquire Project Team

See 9.2 of the *PMBOK® Guide*—Third Edition, which is applicable to construction projects, although these projects may include additional factors. Frequently construction staff is from the contractor's or engineer/contractor's permanent organization, supplemented by local hires, where possible. However, in the case of a joint venture, the staff may be a mixture of personnel from the sponsoring firm and other member firms of the joint venture. Sometimes a construction firm may be asked to provide personnel to an integrated organization, consisting of its own staff and that of the owner.

9.2.1 Acquire Project Team: Inputs

See Section 9.2.1 of the *PMBOK® Guide*—Third Edition, with modifications to Sections 9.2.1.1 and 9.2.1.5, as follows:

.1 Enterprise Environmental Factors

See also Section 9.2.1.1 of the *PMBOK® Guide*—Third Edition. In addition to the factors listed, three other factors should be considered in construction:

- **Recruitment Practices.** Local governments and certain funding agencies may require the use of a local labor workforce (or percentage thereof). Such a defined workforce may not possess the skills required to perform the required tasks either from a quality or productivity standpoint.
- **Local Work Culture.** The work culture, practices, and ethics of the local workforce must be known in advance. Productivity, workmanship, and commitment can vary dramatically from culture to culture around the globe. Such factors may affect the composition of the teams and number of workers required for a specific task. In addition to the work culture, the local language is a consideration, often requiring bilingual project staff or interpreters to communicate across the project team and construction trades.
- **Collective Agreements.** Union agreements must be reviewed with respect to renewal dates, payment provisions, and workplace restrictions. The management team should take precautionary steps in the event of work stoppages resulting from unsuccessful collective bargaining or other union contract violations.

.5 Staffing Management Plan

See also Section 9.2.1.5 of the *PMBOK® Guide*—Third Edition. In addition to the factors mentioned in the *PMBOK® Guide*—Third Edition, construction projects also need to consider a staffing pool description, including personal constraints such as family needs and disabilities that could affect willingness to take on an assignment or ability to perform. This is especially relevant in foreign assignments where the ability to make personnel changes is limited.

9.2.2 Acquire Project Team: Tools and Techniques

See 9.2.2 of the *PMBOK® Guide*—Third Edition, with additional construction-related information for 9.2.2.1 through 9.2.2 4:

.1 Pre-Assignment

See also Section 9.2.2.1 of the *PMBOK® Guide*—Third Edition. Managerial staff may sometimes be pre-assigned to the project. This is often the case when the project is the result of a competitive proposal and specific staff assignments are promised as part of the proposal.

.2 **Negotiation**

See also Section 9.2.2.2 of the *PMBOK® Guide*—Third Edition, which states, "Staff assignments are negotiated on many projects." Since, in many cases, the staff for a construction project may come from a company's organization, negotiations take place between the project personnel and other divisions or departments to obtain needed personnel. In addition, other construction projects may compete for personnel. Negotiations with prospective personnel may be related to the constraints mentioned in Staffing Management Plan (Section 9.2.1.5 of this extension), as well as salary, fringe benefits, responsibilities, and related factors. Such negotiations may involve either current personnel or new hires.

A union trades workforce is usually obtained from the local union hiring hall. The contractor can negotiate with the union on the number and type of workers as well as pay scale and benefits on larger projects (usually through the collective bargaining process). The results of the negotiation can vary depending upon the state of the construction economy and employment level among union members. For non-union projects, it is usually the contractor's responsibility to acquire the workforce locally or from the contractor's own workforce database. In cases of foreign projects or where a large labor force is required, the contractor may be forced to negotiate agreements with local firms that essentially control the available workforce.

.3 **Acquisition**

See also Section 9.2.2.3 of the *PMBOK® Guide*—Third Edition. While the process of recruiting new personnel for a construction staff is a "procurement" process, it is usually performed by a company's human resource or personnel department. When the project hires local personnel to augment the site staff, this recruitment is usually performed by the site staff individual assigned to the human resource responsibility.

.4 **Virtual Teams**

See also Section 9.2.2.4 of the *PMBOK® Guide*—Third Edition. Construction projects and the organizations responsible for these projects are very familiar with the virtual team environment. Most are well experienced with the use of a construction site team and the home office support team. The technological advancements have greatly enhanced the ability of both the design and construction teams in conveying project information, resolving problems, and shortening the turn-around time for clarifications. The ability to issue design changes on a set of drawings, send them electronically to the site, and resolve clarifications immediately, are a few of the many tasks performed by these virtual teams.

9.2.3 Acquire Project Team: Outputs

See Section 9.2.3 of the *PMBOK® Guide*—Third Edition, with the addition of Section 9.2.3.4 as follows:

.4 **Contract Reporting Requirements**

The contract documents often specify the type and frequency of human resource reporting as it pertains to certified payrolls, local beneficial use requirements for M/W/ESB reporting, and, with new international security requirements, the identification of nationalities, site staffing, and regulated controls.

9.3 Develop Project Team

See Section 9.3 of the *PMBOK® Guide*—Third Edition, which states "Develop Project Team improves the competencies and interaction of team members to enhance project performance."

For other types of projects, individual team members are accountable to both a functional manager and the project manager, although this is not always the case on a construction project because of its relative autonomy. There may be an authority higher than the construction manager on a construction jobsite who oversees the safety and quality functions.

Typical construction projects have a duration of six months to two years and, in some cases, longer. Within that timeframe, several components requiring different teams of the labor force may be required. For example, the concrete foundation crew is comprised of different personnel from the crew involved in final fitting up (fit-up) and fixtures. It is often difficult, therefore, to apply team development strategies in the traditional sense to the labor workforce. Once each specialized team is identified, it is difficult to change this structure within the context of a project without affecting the project's momentum. Team development, as described in the *PMBOK® Guide*—Third Edition, can, however, be applied as a global effort within an organization over a series of projects.

9.3.1 Develop Project Team: Inputs

See Section 9.3.1 of the *PMBOK® Guide*—Third Edition, with the addition of Sections 9.3.1.4 and 9.3.1.5 as follows:

.4 Performance Reports

Performance reports (Section 10.3.3.1 of the *PMBOK® Guide*—Third Edition) form the basis for evaluation of individual performance, which become a part of the permanent personnel records and provide support for salary increases and promotions. On projects with multiple repetitive tasks, the labor workforce can work with the management team to continuously improve efficiency, quality, and performance over a period of time. Where the labor force is largely furnished by subcontractors, the subcontractor's overall performance can be evaluated and noted for future use. Performance reports can also be used to document the project team's performance in the areas of timely response, change-order processing, decision making, and construction crew production. See also Performance Reporting (Section 10.3.1 of the *PMBOK® Guide*—Third Edition).

.5 External Feedback

External feedback for a construction project may include contract goals in addition to the usual schedule and cost targets. For example, in a cost-incentive-fee contract, the team's performance is measured periodically (perhaps quarterly) on other items such as quality, safety, communication effectiveness, etc., and that rating determines a portion of the contractor's fee for the period.

9.3.2 Develop Project Team: Tools and Techniques

See Section 9.3.2 of the *PMBOK® Guide*—Third Edition, with modifications to Sections 9.3.2.2 through 9.3.2.6 as follows:

.2 Training

See also 9.3.2.2 of the *PMBOK® Guide*—Third Edition, which discusses the need for and source of training. Training for team members assigned or about to be assigned to a construction project can be more difficult than for other types of projects because many field assignments are not known far enough in advance to provide such training. In addition to the areas discussed in the *PMBOK® Guide*—Third Edition, there other areas where skills can be supplemented on the job, such as safety and quality. It is useful to present a seminar to the team so that they understand the project (for example, in the case of an industrial project, it would be ideal if appropriate members of the owner's operating team were involved. Refer to Section 9.2.3.4 of this extension for potential contract-reporting requirements related to training, such as On-the-Job Training (OJT) and various trade apprentice reporting.

.3 Team-Building Activities

See Section 9.3.2.3 of the *PMBOK® Guide*—Third Edition. Construction management often schedules sports contests, holiday outings, picnics, and dinners to provide an environment that promotes team building and awareness. Other traditional team-building activities are difficult to implement on construction projects, due to their limited duration and varied total team composition (as mentioned in the introduction to this section). Refer to Section 16.3.2.5 in this extension on partnering as a technique for developing the integrated organizational project team.

.4 Ground Rules

See Section 9.3.2.4 of the *PMBOK® Guide*—Third Edition. Also mentioned in Section 16.3.2.5 of this extension, is a technique of partnering to establish the ground rules for integrated project teams, including issue escalation, authority levels, and decision making at the site level.

.5 Co-Location

See also Section 9.3.2.5 of the *PMBOK® Guide*—Third Edition. For a construction project, generally all members of the team are co-located except for the designer, who is an important member of the project team. On very large projects, only a portion of the design team may be onsite and, even then, it is usually for a limited time. In this case, emphasis must be placed on effective communication (see Section 10 of the *PMBOK® Guide*—Third Edition) regardless of whether the constructing organization is an engineer/contractor or not. With the increasing number of design/build projects, designers are experiencing the co-location of their engineers and designers as well.

.6 Recognition and Rewards

See also Section 9.3.2.6 of the *PMBOK® Guide*—Third Edition, which is also applicable to construction projects. Reward for excellent safety performance by the construction team is another type of reward and recognition tool and technique used in this industry. For meeting or exceeding the project's safety goals, the staff and, on occasion, the workers are rewarded with a ceremonial lunch or dinner and are recognized in some way for the achievement. In some cases, this is personal merchandise such as a cap or jacket or participation in lotteries for major prize items.

9.3.3 Develop Project Team: Outputs

See Section 9.3.3.1 of the *PMBOK® Guide*—Third Edition.

9.4 Manage Project Team

See Section 9.4 in the *PMBOK® Guide*—Third Edition.

9.4.1 Manage Project Team: Inputs

See Section 9.4.1 of the *PMBOK® Guide*—Third Edition.

9.4.2 Manage Project Team: Tools and Techniques

See Section 9.4.2 of the *PMBOK® Guide*—Third Edition, with modification of Section 9.4.2.4, as follows:

.4 Issue Log

The timeliness and resolution of issues on a construction project are essential to the progressive flow of work sequence and work performance of the performing organizations. The use of an issues log must be administered by the project team. As described in Section 16.3.2.5 of this extension, Project Partnering is a technique that can be established to enhance the interactions among the stakeholders at the site level and between levels of the project team. Recognizing the roles and responsibilities of personnel is essential for the decision-making authority. Refer to Roles and Responsibilities (Section 9.2.1.3 of the *PMBOK® Guide*—Third Edition).

9.4.3 Manage Project Team: Outputs

See Section 9.4.3 of the *PMBOK® Guide*—Third Edition.

9.5 Close Project Team

The closeout and dissolution of the project team is a major consideration for construction projects. Usually an entire team is not together by the end of the project. As certain responsibilities are completed, team members are released and either *(a)* return to their source department, *(b)* are assigned to another (construction) project, or *(c)* are returned to their point of hire and the open job market. When the project is completed, all of the remaining team members, including the project/construction manager are released from the project. As a project approaches its conclusion, some members will take steps to find subsequent employment or may delay the completion of their assignment, either of which can cause considerable difficulty for the construction manager, if it is not anticipated and addressed. See Incentives (Section 9.5.2.3 of this extension).

9.5.1 Close Project Team: Inputs

.1 **Staffing Management Plan**
See Section 9.1.3.3 of this extension.

.2 **Project Management Plan**
See Section 4.3.1 on Project Plan Development in this extension.

.3 **Personal Contract Requirements**
Occasionally for foreign construction assignments, staff personnel are engaged through the use of personal contracts that specify items, such as salary, living accommodations, tax treatment, and repatriation terms. At project closeout, these contract requirements need to be honored and may require extra attention from the project/construction manager and the staff member assigned to the human resource responsibility.

.4 **Company Policies and Procedures**
For others, although there may not be a personal contract involved, company policy may dictate how personnel are to be released from the project and what procedures are to be followed.

9.5.2 Close Project Team: Tools and Techniques

.1 **Personal Exit Interviews**
As a project approaches completion and personnel are released, it is a good practice to hold an exit interview with each staff member. During this interview, information can be obtained about ongoing activities that may not have been completed, relevant claim documentation, and other data that the project may need, including a discussion about the function of the project team and what lessons have been learned that may improve future performance. See also Lessons Learned Summary Report (Section 9.5.3.3 of this extension).

.2 **Personnel Evaluations**
If periodic performance evaluations are not done, they should be performed during the closeout period. Personnel deserve to know how well they have performed and what kind of recommendation they will receive as a result of their work on the project.

.3 **Incentives**
In some cases, staff members may foresee the close of their assignment and initiate steps to seek employment elsewhere, which can leave the project seriously short-handed in the critical closing phases. One way to offset this problem is to offer a bonus for staying until the final close of the project. Obviously, the project/construction manager has to be careful in the use of this tactic, otherwise everyone could claim they are qualified for such a bonus. Nevertheless, this technique can be useful if such a situation arises.

9.5.3 Close Project Team: Outputs

.1 Project Team Demobilized

When the project is completed, the team leaves the site and future disposition is completed for each member.

.2 Personnel Records

At the close of the project, all personnel records should be reviewed for completeness and forwarded to the company's Human Resource Department, which normally maintains a permanent file on all employees.

.3 Lessons Learned Summary Report

As the project nears completion, the project team will be downsized, along with other personnel. This is the best time to capture the lessons learned from the project personnel in terms of what went well and which processes and system areas could be improved. Gathering this type of information and summarizing it, not only from your organization, but from other participating organizations, can greatly enhance future project efficiencies and successes. It also demonstrates to the stakeholders your intent to deliver high quality project management consistently.

Chapter 10

Project Communications Management

See Chapter 10 of the *PMBOK® Guide*—Third Edition, which states, "Project Communications Management is the Knowledge Area that employs the processes required to ensure timely and appropriate generation, collection, distribution, storage, retrieval, and ultimate disposition of project information." Perhaps no process is more important in the design and construction of a project because of the number and diversity of key players, and because it is vital to the success of the project that the communication of information be timely and accurate. Consequently, considerable thought and planning is required to provide a system that meets these two criteria. For the construction project, the project team and primarily the project manager are the focal point of project communication, not just for distributing project-generated information, but also for gathering, analyzing and responding to stakeholder-initiated information (feedback).

Project documentation is a major consideration for construction projects due to the extensive design and construction contract requirements for managing the contemporaneous documentation produced during the course of a project. For managing documentation to meet contract and project requirements for quality, contractual administration and subsequent acceptance and operational use of the constructed project, refer also to Section 12.6.1 on Contract Closure: Inputs of the *PMBOK® Guide*—Third Edition. See Figure 10-1 for an overview of the processes and respective inputs, tools and techniques, and outputs.

10.1 Communications Planning

See Section 10.1 of the *PMBOK® Guide*—Third Edition which states, "The communications planning process determines the information and communications needs of the stakeholders; for example, who needs what information, when they will need it, how it will be presented and by whom. While all projects share the need to communicate project information, the informational needs and the methods of distribution vary widely. Identifying the informational needs of the stakeholders and determining a suitable means of meeting those needs is an important factor for project success." This is especially true in design and construction projects, as most communications

PROJECT COMMUNICATIONS MANAGEMENT

10.1 Communications Planning

.1 Inputs
- .1 Enterprise environmental factors
- .2 Organizational process assets
- *.3 Project scope statement*
- .4 Project management plan
  - Constraints
  - Assumptions
- ***.5 Contract documents***

.2 Tools and Techniques
- *.1 Communications requirements analysis*
- .2 Communications technology
- ***.3 Project documentation assessment***

.3 Outputs
- .1 Communications management plan
- ***.2 Project documentation requirements checklist***

10.2 Information Distribution

.1 Inputs
- .1 Communications management plan
- ***.2 Project documentation requirements checklist***

.2 Tools and Techniques
- *.1 Communications skills*
- .2 Information gathering and retrieval systems
- *.3 Information distribution methods*
- .4 Lessons learned process
- ***.5 Manage project documentation***

.3 Outputs
- *.1 Organizational process assets (updates)*
- .2 Requested changes

10.3 Performance Reporting

.1 Inputs
- *.1 Work performance information*
- .2 Performance measurements
- .3 Forecasted completion
- .4 Quality control measurements
- .5 Project management plan
  - Performance measurement baseline
- .6 Approved change requests
- .7 Deliverables

.2 Tools and Techniques
- .1 Information presentation tools
- *.2 Performance information gathering and compilation*
- .3 Status review meetings
- .4 Time reporting systems
- .5 Cost reporting systems
- ***.6 Other reporting systems***

.3 Outputs
- *.1 Performance reports*
- *.2 Forecasts*
- .3 Requested changes
- .4 Recommended corrective actions
- *.5 Organizational process assets (updates)*

10.4 Manage Stakeholders

.1 Inputs
- .1 Communications management plan
- .2 Organizational process assets

.2 Tools and Techniques
- .1 Communications methods
- .2 Issue logs

.3 Outputs
- .1 Resolved issues
- .2 Approved change requests
- .3 Approved corrective actions
- .4 Organizational process assets (updates)
- *.5 Project management plan (updates)*
- ***.6 Stakeholder management plan***

Figure 10-1. Project Communications Management Overview

planning is performed as part of the earliest project phases, particularly the development of the project plan (see Section 4.3 of the *PMBOK® Guide*—Third Edition). During pre-construction services the collaborative process between stakeholders is the primary means to effective version control of construction documents and the collaboration between the contributing entities on the project, including the designer, owner, subconsultants and contractors. Their open communication is crucial to properly launch construction with proper documents and baselines in place.

One of the major considerations in planning a communications system for construction projects is to determine how requests for information (RFI) are to be handled. These communications between the constructor and the designer/project manager can have a serious effect on the cost and schedule of a project and can lead to claims. Although it is practically impossible to estimate how many RFIs there might be over the life of the project, it is important to establish and provide an efficient, quick way of answering them and recording the results. Sometimes these guidelines are, or can be established, in the project contract including the time allowed for answering them. See Sections 16.3.2.4 and 16.3.2.5 of this extension, which describe RFI procedures and emphasize the application of partnering and the alignment of the project management systems and processes as a technique to enhance the coordination and collaboration between the project stakeholders, respectively.

Another important matter relates to the how information is to be conveyed to the various project stakeholders, and therefore, what information is required to be conveyed. It is imperative that when developing the many formats that will be used to record project information, the correct data is identified and captured. This becomes even more critical when different stakeholders use, or indeed need, the captured information for different purposes, for example, the same information could be used for: (1) verifying compliance (or non-compliance) with workmanship/material acceptance criteria; (2) progress reporting; and (3) claims. When developing any record format, it is important to ascertain, precisely, what information is to be conveyed to the target audience, as arbitrarily capturing just any data can be just as poor and lead to similar problems as failing to capture any data at all.

10.1.1 Communications Planning: Inputs

See Section 10.1.1 of the *PMBOK® Guide*—Third Edition, with the modification of Section 10.1.1.3 and the addition of Section 10.1.1.5 as follows:

.3 Project Scope Statement

See also Section 10.1.1.3 of the *PMBOK® Guide*—Third Edition. One item in Section 5.2.3.1 of the *PMBOK® Guide*—Third Edition is constraints, which states, "When a project is performed under contract, contractual provisions will generally be constraints." Such constraints often impact communications planning. This is almost always true for construction projects. In addition, the evaluation of these constraints may be important to assess the distribution of sensitive information only to those who have a need to know. One of the important restrictions applies to those who are authorized to make changes, which is particularly relevant in the use of computer-generated and shared design. Another important required constraint involving change requests is the need for an agreed communication path for those requests to avoid the practice of some owners to (Initiate) changes to unauthorized construction personnel.

.5 Contract Documents

The contract documents, which include the contract, the general and special conditions, and the design documents with their referenced documents, indicate specific record and contemporaneous documentation and reporting requirements. These items need to be generated and managed throughout the duration of the project.

In addition to these inputs, see Project Integration Management (Chapter 4), and its seven major processes, which serve to integrate the extensive overlap and cross-utilization of inputs and outputs of the construction industry.

10.1.2 Communications Planning: Tools and Techniques

See Section 10.1.2 of the *PMBOK® Guide*—Third Edition, with the modification of Section 10.1.2.1 and the addition of Section 10.1.2.3, as follows:

.1 Communications Requirements Analysis

See also Section 10.1.2.1 of the *PMBOK® Guide*—Third Edition, which states, "The analysis of the communications requirements results in the sum of the information needs of the project stakeholders. These requirements are defined by combining the type and format of information needed with an analysis of the value of that information. Project resources are expended only on communicating information that contributes to success or where a lack of communication can lead to failure." In design and construction projects, all team members are not commonly co-located, which places greater demands on the communications system.

Stakeholders for construction projects, beside the obvious ones of the customer and the design and construction team, may include utilities, government agencies, financial institutions, the general public and others who have an interest in or are affected by the project. The communications assessment must include the identification of any regulatory or statutory communications requirements that may be required or impact the project.

.3 Project Documentation Assessment

The requirements of the contract documents determine what must be documented in order to fulfill contractual obligations. However, the extensiveness and level of detail, along with stakeholder internal documentation requirements, needs to be assessed at the start of the project. Often, stakeholders assume that everything will proceed well on the project, until something goes wrong. At this point, it is important to have documentation to support their position and illustrate the real situation at hand. Specific documents, such as the daily field reports, inspection reports, submittal approval, jobsite visitors log, telephone, conversation record or a well-described RFI with project delays noted, are just a few examples. These types of documents and the level of detail become extremely important in the case of a dispute. Refer to Chapter 16 on Project Claim Management, for a further discussion of adequate site and project documentation. Well-established construction stakeholders frequently have jobsite and project communication procedures in place. These should be evaluated in relation to a project's size and complexity, and should not preclude any of the required basic documentation. A sample set of project documents, which support the stakeholders and project requirements, can set the precedent for adequate and thorough documentation.

10.1.3 Communications Planning: Outputs

See Section 10.1.3.1 of the *PMBOK® Guide*—Third Edition, with the addition of Section 10.1.3.2 as follows:

.2 Project Documentation Requirements Checklist

The communication management plan should include the project documentation requirements, however, the construction industry has many unique requirements involving labor, jurisdictional reporting, certified payrolls, and other regulations that are designed to protect the workers, environment, and enterprise environmental factors. A checklist of these documentation requirements can serve as a supplement to the communication management plan. Refer to Sections 10.1 and 10.3 of this extension for discussions on the possible types of labor reports and documentation.

10.2 Information Distribution

See Section 10.2 of the *PMBOK® Guide*—Third Edition, which states, "Information Distribution involves making information available to project stakeholders in a timely manner. Information distribution includes implementing the communications management plan, as well as responding to unexpected requests for information."

10.2.1 Information Distribution: Inputs

See Section 10.2.1 of the the *PMBOK® Guide*—Third Edition, with the addition of Section 10.2.1.2 as follows:

.2 Project Documentation Requirements Checklist

A construction project has many different participants who work in a remote location away from the project site. Often, these home office operations produce and distribute a variety of reports and regulated documentation. These project participants do not utilize specific project communication plans, but rather are required to adhere to policy, jurisdictional or international regulations, employee tax and welfare payments, for example, any documentation that has regulated frequency and distribution requirements (See also Section 10.1.3.2 of this extension).

10.2.2 Information Distribution: Tools and Techniques

See 10.2.2 of the *PMBOK® Guide*—Third Edition, with modifications to 10.2.2.1 and 10.2.2.3, and an additional construction-related tool and technique in 10.2.2.5.

.1 Communications Skills

See also Section 10.2.2.1 of The *PMBOK® Guide*—Third Edition, which states, "Communications skills are part of general management skills and are used to exchange information. General management skills related to communication include ensuring that the right persons get the right information at the right time, as defined in the communications management plan. General management skills also include the art of managing stakeholder requirements."

In construction-related projects, it is possible that some skills do not exist within the project team. It may be required to obtain specific professional services, such as a public relations consultant.

.3 Information Distribution Methods

See also Section 10.2.2.3 in the *PMBOK® Guide*—Third Edition. Project information may be distributed using a variety of methods, including kick-off and regular project meetings, hard-copy document distribution, shared access to networked electronic databases, fax, electronic mail, voice mail, videoconferencing, and project intranet. For some larger projects a central communications department could be established as a means of distributing project information to those heavily involved with the project. In addition, the use of public relations or other specialty services may be required.

Much of the documentation generated by the construction project is both time sensitive and approval- based, such as shop drawings, changes, and design clarification requests. The use of project logs facilitates the continual exchange, turn-around times, approvals, and the assigned document number corresponding to the type of documentation. These logs become an essential part of the project records and a record source for recording dates and instructions of contemporaneous documentation. See also Section 16.1.2.3 on Project Claim Management in this extension.

.5 Manage Project Documentation

The vast amount and extensive variety of project documentation on a construction project can be overwhelming if not managed effectively from the onset. Effective and efficient administration of the documentation is critical and must be integrated throughout the life of the project. From contracts to pay estimates, from design clarifications to change orders, from performance reporting to owner/maintenance manuals and warranties, the construction documentation is voluminous. Consequently, the communication plan and the staff charged with administering it may require the additional capability, systems, and processes to integrate and manage the volume of individual documents and the paper flow. Often, each stakeholder may have its own ritual of organizing the project records; however, a consistent document file structure is the preferred technique. The following items should be considered as additional communication plan components which address the integration aspects and importance of the various types of documentation:

- **Manage system administration and records**, including project systems planning, requirements identification, system monitoring and control, system process improvement planning and implementation.
- **Manage and archive performance measurement documentation**, including the planning documents that established the framework for performance measurement, the project contemporaneous records, the project as-built .conditions in the form of drawings, process diagrams, equipment manuals, operational and training specifications/manuals.
- **Manage and retain the project archives**, including the project management plan (updates), integrated environmental, safety, security, financial and quality plans. Refer also to Section 4.7 of the *PMBOK® Guide*—Third Edition for additional requirements.

10.2.3 Information Distribution: Outputs

See Sections 10.2.3 of the *PMBOK® Guide*—Third Edition, with the modification to Section 10.2.3.1 as follows:

.1 Organizational Process Assets (Updates)

See Section 10.2.3.1 of the *PMBOK® Guide*—Third Edition and add the following:

- **Project Logs.** The project team maintains the logs by shared use of the specific electronic files established for the project. This type of output allows one to quickly retrieve pertinent data regarding issues and records without having to review multiple and extensive files containing the actual documents.

10.3 Performance Reporting

See Section 10.3 of the *PMBOK® Guide*—Third Edition, which states, "The performance reporting process involves the collection of all baseline data, and distribution of performance information to stakeholders. Generally this information includes how resources are being used to achieve project objectives."

Construction projects usually also require information on risk and procurement. Reports may be prepared comprehensively or on an exception basis. Information distribution in a design/construction project is important enough to warrant a separate section for reporting. Examples of progress performance techniques are detailed in Section 10.3.3.1 of this extension.

10.3.1 Performance Reporting: Inputs

See Section 10.3.1 of *PMBOK® Guide*—Third Edition, with the modification of Section 10.3.1.1 as follows:

.1 Work Performance Information

See Section 10.3.1.1 of the *PMBOK® Guide*—Third Edition. Due to the potential for interpretation arguments and contractual disputes on construction projects, the importance of contemporaneous documentation is very high and remains a priority throughout the life of the project. See also Information Gathering and Retrieval Systems (Section 10.2.2.2 of the *PMBOK® Guide*—Third Edition) for further emphasis on information gathering and distribution. This type of documentation is fundamental to resolving disputes and adequately describing circumstances.

10.3.2 Performance Reporting: Tools and Techniques

See Section 10.3.2 of the *PMBOK® Guide*—Third Edition, with the modification of Section 10.3.2.2 and the addition of Section 10.2.3.6 as follows:

.2 Performance Information Gathering and Compilation

See also Section 10.3.2.2 of the *PMBOK® Guide*—Third Edition. Again the importance of contemporaneous documentation is noted to support its critical nature involving dispute resolution. The means and methods for managing this type of information are critical.

.6 **Other Reporting Systems**

Among the safety, environmental, financial, and extensive quality control requirements that exist in the construction industry, many have specialized functions and contain specific reporting systems based on the information and documentation to be gathered and recorded. It is important to note that the tools and techniques required for these specialized areas may vary, but are an essential component of the performance reporting. Refer also to Chapters 8, 13, 14, and 15 of this extension for the requirements of inputs, tools and techniques, and outputs.

10.3.3 Performance Reporting: Outputs

See Section 10.3.3 of the *PMBOK® Guide*—Third Edition, with the modification of Sections 10.3.3.1, 10.3.3.2, and 10.3.3.5 as follows:

.1 **Performance Reports**

See also Section 10.3.3.1 of the *PMBOK® Guide*—Third Edition, which states "Performance reports organize and summarize the information gathered and present the results of any analysis as compared to the performance measurement baseline. Reports should provide the status and progress information and the level of detail required by various stakeholders, as documented in the communications management plan."

Performance reports include periodic (often monthly) project status reports that describe the status of the project, a forecast of future activity of cost and earned value, and a status of design, procurement, expediting, risk evaluation and quality activities of the project. Performance reporting should also include the RFI response record.

Examples of common formats for performance reports include bar charts (also called Gantt charts), S-curves, histograms, and tables. See Project Time Management (Chapter 6 of the *PMBOK® Guide*—Third Edition) for various types of schedules and methods for monitoring and reporting project status and progress. In addition, progress reports for construction projects commonly require the inclusion of information relating to performance measurement analyses such as those detailed in 7.3.2.2 of the *PMBOK® Guide*—Third Edition, to determine the magnitude of variances that can, and often do, occur on construction projects. The main information pertaining to required performance measurement analyses includes:

- Earned value techniques (EVT), in particular Cost Performance Index (CPI) and Schedule Performance Index (SPI) which would be presented in both tabular and graphical form to identify trends in progress (or lack thereof). Indeed, such presentations are often considered a vital element of any construction progress report. Figure 10-2 and 10-3 provide summaries of CPI and SPI information for a 12 month project, with the information presented in graphical form in Figure 10-4 (by Radial Method) and Figure 10-5 (by Linear Method). It should be noted that both Figures 10-4 and 10-5 present the identical information.
- Forecasting, for example, estimate to complete (ETC), and estimate at completion (EAC). See also Section 10.3.3.2 of this extension

  In summary, the project finished both behind schedule and over budget, as from July to October, cost spiraled (CPI fell from 1.055 to 0.980 over this 3–4 month period) but the project progressed ahead of schedule for the same

	Period	Jan	Feb	March	April	May	June	July	Aug	Sept	Oct	Nov	Dec	Overall Average per Discipline
Road Pavement Works	SPI	1.004	1.000	1.029	0.993	0.980	1.006	1.012	1.045	1.039	1.062	1.018	0.962	1.013
	CPI	1.010	0.995	0.989	1.025	1.020	1.028	1.045	0.991	1.010	0.992	0.986	0.976	1.006
Drainage Works	SPI	1.010	1.025	1.026	1.000	0.995	0.985	1.018	0.990	1.065	1.073	1.025	0.983	1.016
	CPI	1.000	1.030	1.050	1.040	1.036	1.034	1.049	1.058	1.036	0.933	0.976	0.950	1.016
Earthworks	SPI	0.995	1.030	1.056	1.050	0.978	0.970	1.009	1.099	0.993	1.050	1.038	0.994	1.022
	CPI	1.005	1.021	1.002	1.035	1.010	1.055	1.070	1.075	0.992	0.994	0.981	0.997	1.020
Bridges and Structures	SPI	1.020	1.025	1.029	1.036	0.985	0.960	1.039	1.065	1.082	1.055	1.039	0.999	1.028
	CPI	1.005	1.015	1.000	1.039	1.014	1.082	1.055	1.017	1.000	0.999	0.996	0.998	1.018
Utilities (Lighting, Services, Barriers, etc.)	SPI	1.021	1.021	1.036	1.029	0.989	0.981	1.021	1.052	1.046	1.061	1.031	0.987	1.023
	CPI	1.004	1.015	1.007	1.034	1.021	1.051	1.058	1.034	1.014	0.982	0.986	0.978	1.015

Figure 10-2. Tabulation of CPI vs SPI per Discipline

	Period	Jan	Feb	March	April	May	June	July	Aug	Sept	Oct	Nov	Dec	Overall Average
Overall	SPI	1.010	1.020	1.035	1.020	0.985	0.980	1.020	1.050	1.045	1.060	1.030	0.985	1.020
	CPI	1.005	1.015	1.010	1.035	1.020	1.050	1.055	1.035	1.010	0.980	0.985	0.980	1.015

Figure 10-3. Tabulation of Averaged CPI vs SPI

Note: The following convention for Figures 10-2 and 10-3 has been used:

Value = Under Budget (CVI > 1.0) or Ahead of Schedule (SVI > 1.0)

Value = Over Budget (CVI < 1.0) or Behind Schedule (SVI < 1.0)

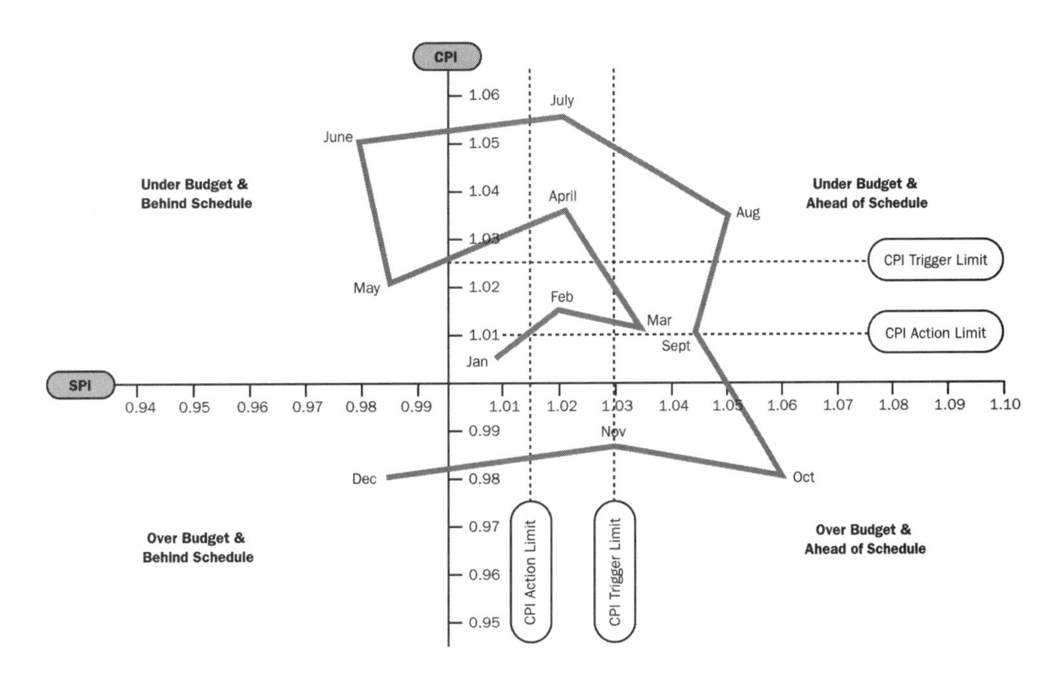

Figure 10-4. Overall Project Performance - from CPI and SPI values (Radial Method)

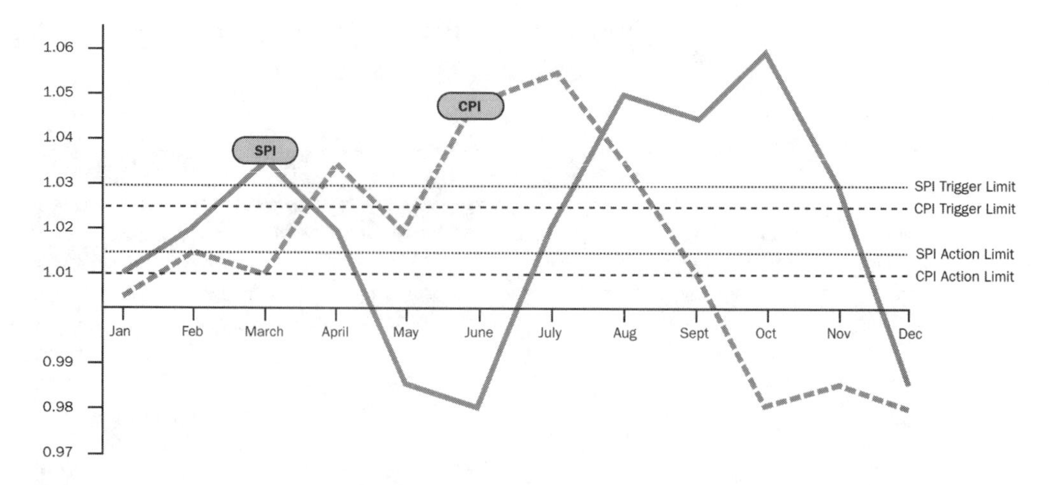

Figure 10-5. Overall Project Performance - from CPI and SPI values (Linear Method)

period. The period from October to December where SPI fell from 1.060 to 0.985 is indicative of significant corrective rework needed or delayed work completion

Figures 10-4 and 10-5 also include details of 'limits' to be used by the performing organization (or indeed by a monitoring organization, i.e., the owner or sponsor, etc.) as indicators of when and where action is needed as trends show or indicate decreasing SPI and/or CPI. These limits are termed:

- **Trigger Limits**—where action should be taken to avoid SPI and/or CPI decreasing to a point when cost and schedule overruns are forecast or expected.
- **Action Limits**—where immediate action is necessary as trends indicate the decreasing SPI and/or CPI will soon place the project behind schedule and/or over budget.

The use and application of such limits will vary depending on the maturity of the project management system and culture of the performing organization (or monitoring organization). These limits need to be carefully scrutinized prior to implementation and will require acceptance to ensure workable values are selected, and not values which are difficult or impossible to achieve.

Presenting the information in tabular form (Figures 10-2 and 10-3) in conjunction with graphical form (Figure 10-4 or 10-5) is very useful where projects have numerous simultaneous and/or consecutive execution phases, for example design and build (DB), engineering, procurement, and construction (EPC), etc., as variances (and their magnitude) can be forecasted or immediately and precisely identified. For a complete picture of an entire project either at a specific point in time or at completion, CPI and SPI values would be aggregated or averaged to provide an accurate overview status of the project. The information presented in Figures 10-2 and 10-3, and Figures 10-4 or 10-5, would be fundamental attributes or elements of project performance reports where financial institutions (World Bank, etc.) or government agencies are involved in project funding and/or project monitoring.

.2 **Forecasts** See also Sections 10.3.3.2 and 4.3 of the *PMBOK® Guide*—Third Edition.

.5 **Organizational Process Assets (updates)**

See also Section 10.3.3.5 of the *PMBOK® Guide*—Third Edition. The construction industry is required to maintain records supporting the fulfillment of constructed components in line with the design. Much of the industry is governed by the potential liability associated with design or construction defects, as well as the life of the constructed facility that is specified by design. These quality management reporting systems (see Sections 8.2 and 8.3) are based on prior approvals for materials, as well as in-place constructed components. Likewise, the jurisdictional requirements affecting the environment, safety, human resources and finances of the project all have specific types, formats, frequency and distribution. These requirements are often more extensive than would be the case with a non-construction project undertaken within a private enterprise or for product or IT development. Refer to each specific chapter for characteristics unique to construction.

10.4 Manage Stakeholders

See Section 10.4 in the *PMBOK® Guide*—Third Edition. There is frequently a vast assortment or range of project stakeholders that can be applicable to any one project. Stakeholder needs and requirements vary greatly depending on the involvement in, and how they may influence the project. It is also not uncommon for the needs and requirements of one stakeholder to differ slightly or significantly from those of another. More often than not, the needs and requirements of all project stakeholders take the form of 'information needs and requirements', and it is this information (in the project management context) that needs to be managed. Project stakeholders therefore generally fall into one of the following category groups:

1. *Direct project stakeholders*—those stakeholders directly involved in the execution of the project, and include, but are not limited to the following:
   - Project sponsors,
   - Project owners,
   - Project designers,
   - Contractors, and
   - Material suppliers, etc.

   The needs and requirements of direct project stakeholders will often be detailed in the contract(s), specifications, and, work standards employed on the construction project.

2. *Indirect project stakeholders*—those not directly involved in the execution of the project, but, can have an influence on project execution, and include, but are not limited to the following:
   - Regulatory agencies or authorities, i.e., regarding safety and environmental issues, etc.,
   - Professional associations,
   - General public, including local residents groups,
   - Labor Unions,
   - Local government departments,
   - Media,
   - Lobbyist or petitioner groups,
   - National industry or business representatives and associations,
   - Police and other emergency services, etc.

The needs and requirements of indirect project stakeholder however, are somewhat more difficult to determine, and can be more ambiguous especially when dealing with stakeholders ardently opposed to the execution of the project. The management of all information that can be generated throughout a project is often seen as a vast task, especially when trying to determine which stakeholders needs which piece of information. A common approach to the management of project stakeholders (and their needs and requirements) is the development of a project stakeholder management plan (see 10.4.6 of this extension) which is subset to the project communication management plan (See Chapter 10 of the *PMBOK® Guide*—Third Edition); however, it addresses how specific project stakeholders will be managed, for example, public meetings, newsletters, etc., can be more effective at conveying overall project information and statistics as they reach a larger stakeholder audience more rapidly than one-on-one meetings.

10.4.1 Manage Stakeholders: Inputs

See Section 10.4.1 of the *PMBOK® Guide*—Third Edition.

10.4.2 Manage Stakeholders: Tools and Techniques

See Section 10.4.2 of the *PMBOK® Guide*—Third Edition.

10.4.3 Manage Stakeholders: Outputs

See also Sections 10.4.3 of the *PMBOK® Guide*—Third Edition, with the modification of Section 10.4.3.5 and the addition of Section 10.4.3.6 as follows:

.5 Project Management Plan (Updates)

With the rise of exposed defects within the design and construction industry, these updates provide confirmation that the facilities have been constructed in compliance with all the regulations and the quality-driven design. Additional outputs may be required within this plan, including performance and operational testing results, envelope guarantees, and commissioning reports for the final constructed facility/product in addition to also intermittent reports throughout the project for various constructed components.

.6 Stakeholder Management Plan

Although stakeholder requirements and expectations are included in the communication plan, many construction projects experience very unique circumstances that may require a specific plan developed for the primary purpose of administering to the needs of the various stakeholders, internal and external to the project. For example, specific operational performance that complies with noise restrictions for a neighborhood, or that addresses the concerns of property owners adjacent to the construction development, are two examples of how outside stakeholders can affect the design and progress of the work. Often, a specific stakeholder management plan can incorporate these special considerations that may be only briefly defined by the contract. A plan of this type might include hours of operations, special signage for businesses, and periodic general public meetings to keep the outside stakeholders informed and cooperative.

Chapter 11

Project Risk Management

See Chapter 11 of the *PMBOK® Guide*—Third Edition, which states that "Project Risk Management includes the processes concerned with conducting risk management planning, identification, analysis, responses, and monitoring and control on a project; most of these processes are updated throughout the project. The objectives of Project Risk Management are to increase the probability and impacts of positive events and decrease the probability and impacts of events adverse to project objectives." Construction risk can be managed with these processes with reasonable adequacy. When possible, the risk management process for the construction project should be integrated with the risk management process for the parent project.

See Figure 11-1 for an overview of the processes and respective inputs, tools and techniques, and outputs.

11.1 Risk Management Planning

See Section 11.1 of The *PMBOK® Guide*—Third Edition.

11.1.1 Risk Management Planning: Inputs

See Section 11.1 of the *PMBOK® Guide*—Third Edition, with the modification of 11.1.1.2 and the addition of 11.1.1.5 as follows:

.2 Organizational Process Assets

See Section 11.1.1.2 of the *PMBOK® Guide*—Third Edition. These may include templates for the organization's risk management plan. When the project is performed by a consortium or joint venture, templates from all companies can be considered and a particular template for the project can be developed by the project team. Other assets are the checklist and lessons learned from previous projects and tender and contract documents. Constructability reviews should be part of the risk management process of the parent project that precedes the construction phase or project. The results of these constructability reviews should be inputs to the risk management process for the construction project.

.5 Project Charter

During the bidding phase, the charter can be a request for proposal, invitation for bid, or a similar document which the bidding team will use to guide risk analysis. When a contract is signed, it should be considered, together with the proposal and bidding documentation, as the project charter.

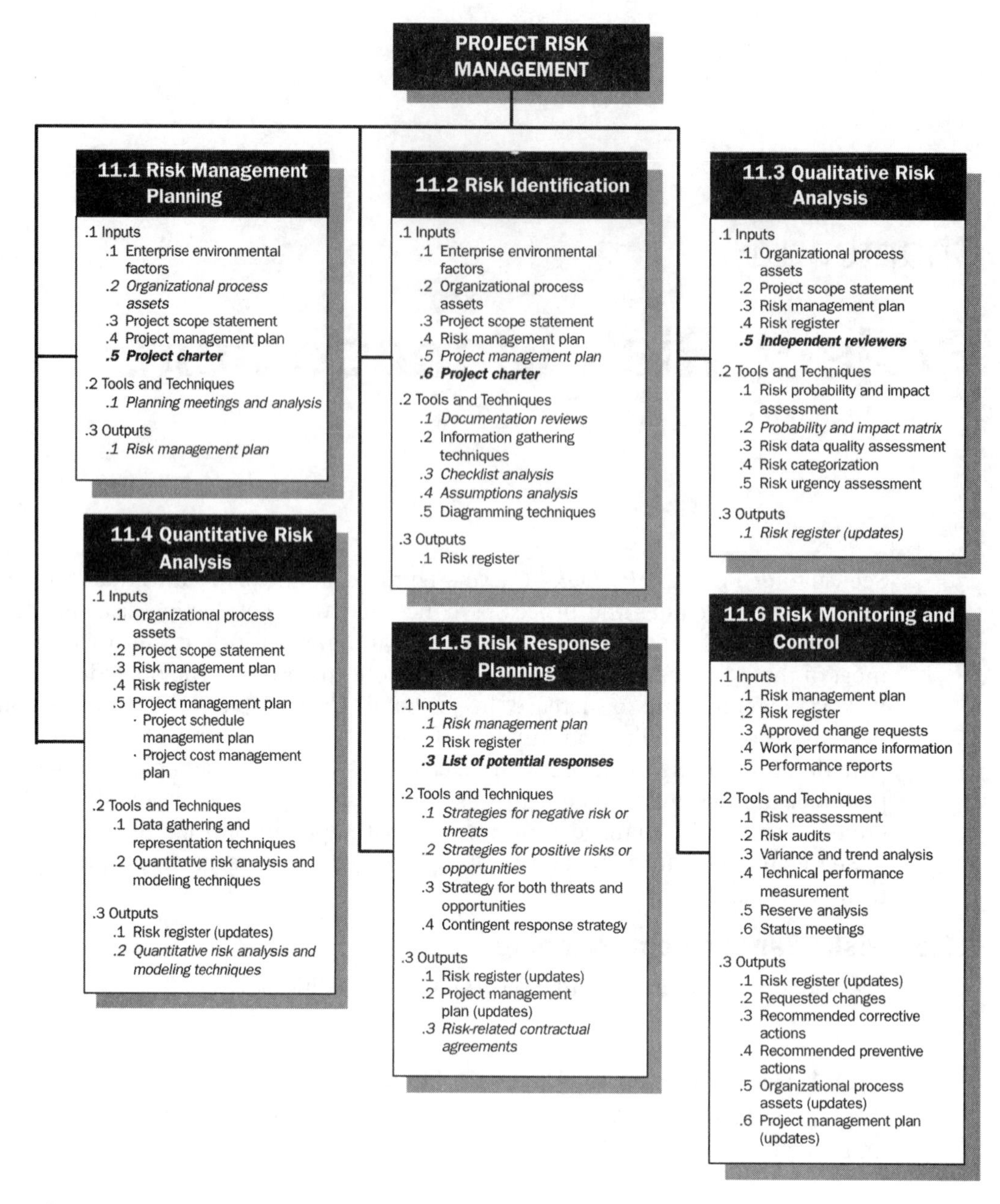

Figure 11-1. Project Risk Management Overview

11.1.2 Risk Management Planning: Tools and Techniques

See Section 11.1.2 of the *PMBOK® Guide*—Third Edition, with the modification of 11.1.2.1, as follows:

.1 Planning Meetings and Analysis

When possible, these meetings should include persons with significant experience in similar projects. The results of these meetings should be shared with the parent project as part of an integrated risk management process for the entire project.

11.1.3 Risk Management Planning: Outputs

See Section 11.1.3 of the *PMBOK® Guide*—Third Edition, with the modification of Section 11.1.3.1 as follows:

.1 Risk Management Plan

The *PMBOK® Guide*—Third Edition describes the risk management plan and the topics it should include. In construction projects, some specific issues unique to construction should be addressed:

- **Methodology**. Addresses topics such as which phases of risk management to perform. It includes individual risk management processes and how their outputs will be linked to the overall project risk management. For example, in an Engineering, Procurement, Construction, and Management (EPCM) project, the engineering and construction phases may have individual risk management processes, while the procurement and management phases can be treated together. Methodology also addresses how safety and environmental risk management plans (Chapters 13 and 14) will interact with the overall construction risk management plan and how the subcontractors' risk management plans will fit into the overall project risk management plan.
- **Budgeting**. Establishes a budget for risk management for the project, which should be equal or less than the amount allocated for that purpose in the bid price.
- **Scoring and interpretation**. For safety and environmental planning, there may be local standards, norms and laws regulating scoring and interpretation methods.

11.2 Risk Identification

See Section 11.2 of the *PMBOK® Guide*—Third Edition, which discusses risk identification and its iterative nature and all iterations which this section suggests apply to construction projects, including a preliminary or initial former iteration that is performed while in the bidding process, to support pricing and contract negotiation decisions. If possible, the parent project should share its assessment of construction risks as input to the bidding process.

11.2.1 Risk Identification: Inputs

See Section 11.2.1 of the *PMBOK® Guide*—Third Edition, with the modification of 11.2.1.5 and the addition of 11.2.1.6 as follows:

.5 Project Management Plan

See Section 11.2.1.5 of the *PMBOK® Guide*—Third Edition. This includes itemization of the categories of risk involved in a given project. In addition to the project Risk Breakdown Structure (RBS) categories illustrated in Figure 11-4 of the *PMBOK® Guide*—Third Edition, some other aspects of risk that are associated with construction are noted as follows:

- **Resource plan**. Resource productivity estimates are a key input to risk identification; activity duration and cost estimates are based on those rates, particularly for engineering and construction phases.
- **Layout risks.** While construction projects are highly dependent on human resources and equipment, adherence to the schedule is a very important

success factor for the project. This is often addressed by resource leveling (Section 6.5.2.5 of the *PMBOK® Guide*—Third Edition), but the facility layout can only be addressed by expert judgment. Formal resource leveling is usually performed on larger projects with a resource-loaded schedule. Informal leveling is most often performed by site management to avoid inefficiency by sequencing the work using schedule logic.

Depending upon the layout designed by engineering, three major problems could arise:

- *Resource overloading.* This is related to the law of diminishing returns. If a productivity rate is achieved with a certain number of resources when more resources are allocated, the tendency is to have a decrease in the rate. The decrease is stronger if the layout is concentrated, with little room between different task fronts;
- *Interference between task fronts.* For example, if the engineering underground systems designers delay their output, construction of those systems can interfere with heavy equipment erection, due to crane positioning or excavation clearances;
- *Project management risk.* This may arise when there is poor coordination of subcontractors' project plans.

.6 Project Charter

The project charter includes the request for the proposal or contract, depending on the project phase. It also may include the product description, since all performance requirements for the facility of which construction is the contract object have to be considered.

11.2.2 Risk Identification: Tools and Techniques

See Section 11.2.2 of the *PMBOK® Guide*—Third Edition, with modifications to Sections 11.2.2.1, 11.2.2.3, and 11.2.2.4 as follows:

.1 Documentation Reviews

In construction projects, besides documents described in the *PMBOK® Guide*—Third Edition, documents such as layout drawings, plant location and access, and equipment erection specifications should be included in the review. Permits, licenses, and agreements with labor unions and/or communities may also include requirements that add risk to the project. All documents produced by the parent project from its constructability reviews should be included in these reviews.

.3 Checklist Analysis

For construction projects, checklists may include such items as: type of contract, unfavorable clauses, site and area factors, weather, regulatory and labor factors, knowledge of client, etc., including the usual appraisal of construction equipment requirements, techniques needed, and special materials.

.4 Assumptions Analysis

For construction projects, all assumptions made during the bidding phase should be reviewed from time to time.

11.2.3 Risk Identification: Outputs

See Section 11.2.3 of the *PMBOK® Guide*—Third Edition.

11.3 Qualitative Risk Analysis

See Section 11.3 of the *PMBOK® Guide*—Third Edition, which states "Qualitative risk analysis includes methods of prioritizing the identified risks for further action such as Quantitative Risk Analysis (Section 11.4) or risk response planning (Section 11.5). Organizations can improve the project's performance effectively by focusing on high-priority risks. Qualitative Risk Analysis assesses the priority of identified risks using their probability of occurring and the corresponding impact on project objectives if the risks do occur, as well as other factors such as the time frame and risk tolerance of the project constraints of cost, schedule, scope, and quality."

11.3.1 Qualitative Risk Analysis: Inputs

See Section 11.3.1. of the *PMBOK® Guide*—Third Edition, with the addition of Section 11.3.1.5 as follows:

.5 Independent Reviewers

Independent reviewers are persons outside the project who have significant experience in similar projects.

11.3.2 Qualitative Risk Analysis: Tools and Techniques

See Section 11.3.2 of the *PMBOK® Guide*—Third Edition, with the modification of Section 11.3.2.2 as follows:

.2 Probability and Impact Matrix

See Section 11.3.2.2 of the *PMBOK® Guide*—Third Edition. For assessing risk probability and impact with the use of a rating matrix, expert judgment is commonly used. In construction projects, care should be taken with the use of expert judgment, as those individuals may represent different stakeholders and different interests, introducing a bias in the risk analysis. For example, experts from the client, main contractor, subcontractors, and government agencies may set quite different values for the probability and impact of a given risk. These differences themselves highlight the risk and its nature.

11.3.3 Qualitative Risk Analysis: Outputs

See Section 11.3.3 of the *PMBOK® Guide*—Third Edition, with the following modification:

.1 Risk Register (Updates)

See Section 11.3.3.1 of the *PMBOK® Guide*—Third Edition, which states, "The risk register is updated with information from Qualitative Risk Analysis and the updated risk register is included in the project management plan (Section 4.3)." Among the several forms of updating itemized in this section is the relative ranking or priority list of project risks. In construction projects, the risk ranking is a useful tool to support the go-no-go decision to bid in response to an RFP.

11.4 Quantitative Risk Analysis

See Section 11.4 of the *PMBOK® Guide*—Third Edition. The Quantitative Risk Analysis process is not common in construction projects, but is very useful to support project management decisions. This process often produces the best information supporting the needs analysis of contingency and management reserve. Quantitative risk analysis uses techniques such as Monte Carlo simulation and decision analysis to determine many project assumptions, for example, as follows:

- Determine the probability of achieving a specific project objective. When bidding, if the final project date is not determined by the client, the project team can offer a date, based on the risk level they are willing to accept.
- Identify realistic and achievable cost, schedule, or scope targets. Acceleration plans benefit from this analysis and aggressive targets can be negotiated with a lower degree of uncertainty.

11.4.1 Quantitative Risk Analysis: Inputs

Section 11.4.1 of the *PMBOK® Guide*—Third Edition discusses inputs to quantitative risk analysis.

11.4.2 Quantitative Risk Analysis: Tools and Techniques

Section 11.4.2 of the *PMBOK® Guide*—Third Edition describes the tools and techniques for quantitative risk analysis.

11.4.3 Quantitative Risk Analysis: Outputs

See Section 11.4.3 of the *PMBOK® Guide*—Third Edition, with the modification of Section 11.4.3.2 as follows:

.2 Quantitative Risk Analysis and Modeling Techniques

The following are additional Quantitative Risk Analysis techniques:

- Cash-flow analysis such as Discounted Cash Flow Return on Investment (DCFROI), Internal Rate of Return (IRR), and Net Present Value (NPV).
- Factorial design of sensitivity analysis.
- Monte Carlo analysis.

11.5 Risk Response Planning

See Section 11.5 of the *PMBOK® Guide*—Third Edition, which states "Risk Response Planning is the process of developing options and determining actions to enhance opportunities and reduce threats to the project's objectives. It follows the qualitative risk analysis and quantitative risk analysis processes. It includes the identification and assignment of one or more persons (the "risk response owner") to take responsibility for each agreed-to and funded risk response." In construction projects, because there is the involvement of subcontractors, Risk Response Planning is a more complex process, as risk responses may result in additional costs incurred by one party to influence outcomes of events that will impact another party. Timing needs to be carefully planned for risk management processes and risk responses are negotiated.

11.5.1 Risk Response Planning: Inputs

See Section 11.5.1 of the *PMBOK® Guide*—Third Edition, with the modification of 11.5.1.1 and the addition of 11.5.1.3 as follows:

.1 Risk Management Plan

See Section 11.5.1.1 of the *PMBOK® Guide*—Third Edition. This plan should record the thresholds, contingencies, and management reserves for the construction project.

.3 List of Potential Responses

See Risk Register (Section 11.5.1.2 of the *PMBOK® Guide*—Third Edition).

11.5.2 Risk Response Planning: Tools and Techniques

See Section 11.5.2 of the *PMBOK® Guide*—Third Edition, with modifications to Sections 11.5.2.1 and 11.5.2.2 as follows:

.1 Strategies for Negative Risks or Threats

See Section 11.5.2.1 of the *PMBOK® Guide*—Third Edition, which lists three strategies: avoid, transfer, and mitigate. The first two of these call for special comment with regard to the construction industry:

- **Avoid**. Risk avoidance is more effective during early project phases and contract negotiation. After a contract is signed, some of the major risks regarding plant performance and penalties cannot be avoided.
- **Transfer.** The *PMBOK® Guide*—Third Edition states, ''Risk transference is seeking to shift the negative impact of a threat to a third party together with ownership of the response. Transferring the risk simply gives another party responsibility for its management; it does not eliminate it.'' The most common form of transference is insurance. The construction industry uses a variety of insurance products to handle some of their liabilities: builder's risk, umbrella policies, and other special types of insurance to cover environmental and currency risks. Subcontracting is also very common in construction projects for many reasons, the greatest of them being the broad scope of construction contracts and the specialization of subcontractors. However, the decision to subcontract is seldom made as a result of risk response planning. A better approach would be to plan risk responses before the subcontract bidding process starts, to achieve a smoother, more effective, transference of risk. Timing is a critical factor in risk management of construction projects.

.2 Strategies for Positive Risks or Opportunities

See Section 11.5.2.2 of the *PMBOK® Guide*—Third Edition. A strategy to share the benefits of a positive risk or opportunity between the project participants should be included in the construction project contract.

11.5.3 Risk Response Planning: Outputs

See Section 11.5.3of the *PMBOK® Guide*—Third Edition, with the modification of 11.5.3.3 as follows:

.3 Risk-Related Contractual Agreements

See Section 11.5.3.3 of the *PMBOK® Guide*—Third Edition. In construction, the *force majeure* is a commonly used phrase, which refers to responsibilities resulting from "acts of God." A variety of clauses can be written to properly divide liabilities between the owner and contractor based on their ability to control them. There are other unknown, unanticipated, and/or hidden risks that should also be addressed. These include underground conditions, weather conditions, commodity and labor pricing changes, and changes from actions external to and independent of the construction project, such as public infrastructure and regulatory actions.

11.6 Risk Monitoring and Control

See Section 11.6 of the *PMBOK® Guide*—Third Edition.

11.6.1 Risk Monitoring and Control: Inputs

See Section 11.6.1 of the *PMBOK® Guide*—Third Edition.

11.6.2 Risk Monitoring and Control: Tools and Techniques

See Section 11.6.2 of the *PMBOK® Guide*—Third Edition.

11.6.3 Risk Monitoring and Control: Outputs

See Section 11.6.3 of the *PMBOK® Guide*—Third Edition.

Chapter 12

Project Procurement Management

See also the introduction to Chapter 12 of the *PMBOK® Guide*—Third Edition. Project Procurement Management (procurement) includes the processes required to acquire construction-related goods and services. With respect to construction, procurement addresses a wide variety of possible deliverables and is also referred to as capital project development. Examples include the construction of a new facility, expansion, renovation or improvement of an existing facility, or the disposition/decommissioning of an outmoded facility.

The *PMBOK® Guide*—Third Edition discusses procurement within the context of the buyer-seller relationship, which exists on many levels in a typical construction project. The buyer may be characterized as a client, customer, prime contractor, governmental agency, etc. Numerous labels may also refer to the seller, including bidder, contractor, subcontractor, vendor, and supplier. See Figure 12-1 for an overview of the processes and respective inputs, tools and techniques, and outputs.

12.1 Plan Purchases and Acquisitions

See also Section 12.1 of *PMBOK® Guide*—Third Edition. The Plan Purchases and Acquisitions process set forth in *PMBOK® Guide*—Third Edition identifies which products, services, or results should be achieved by project members inside the organization and which items should be acquired from outside sellers. Current practice dictates that businesses focus on their own core competencies, and this has resulted in the outsourcing of most construction projects by buyer organizations. This is also true of tangential construction tasks, such as maintenance of facilities. In addition, many seller organizations further down the construction hierarchy are outsourcing activities that were once performed in-house. For example, in the past, general contractors performed a significant amount of the construction work using their own forces and it was common for a general contractor to have masonry, carpentry and concrete crews. The current trend, however, is for the general contractor to focus on managing the work and subcontracting out specific construction activities.

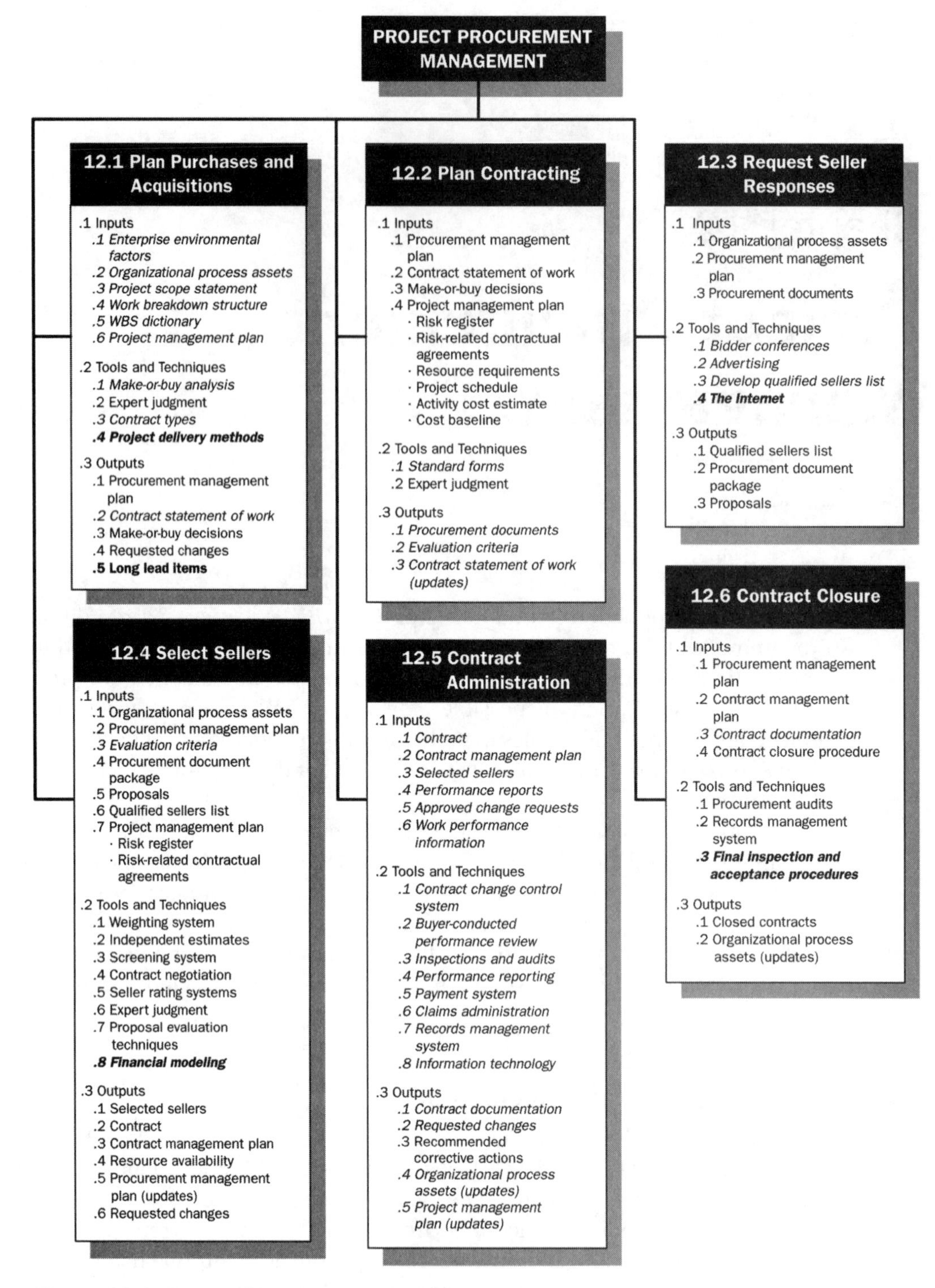

Figure 12-1. Project Procurement Management Overview

From the construction perspective, the acquisitions process addresses the proposed scope of work, the breakdown of the work activities into discrete packages, identification of the project stakeholders, and the timing of their participation.

The Procurement Management Plan of the buyer organization is an important output of the acquisitions process, and is influenced by organizational policies, culture and procedures, and lessons learned from previous construction projects The procurement plan is a critical aspect of the project plan and has a significant influence over how controls of cost, schedule and quality functions will be exercised on contracts. Moreover, the procurement plan reflects senior management's attitude for risk-taking and identifies major categories of risks and their allocation among the stakeholders.

12.1.1 Plan Purchases and Acquisitions: Inputs

See Section 12.1.1 of the *PMBOK® Guide*—Third Edition, with the following modifications:

.1 Enterprise Environmental Factors

See also Sections 12.1.1.1 and 4.1.1.3 of the *PMBOK® Guide*—Third Edition. Organizations may outsource construction-related activities, therefore the availability of management personnel, skilled labor, and material inventory from the sellers in the marketplace are a major concern. Adding to the concern are factors that can significantly impact how procurements are packaged, solicited, awarded, and executed. The factors include location, culture, climate, and political stability, which affect contractor liability and competency, work rules, work hours, labor competency, and productivity.

Material availability may also be of great importance. Even in a global economy, shortages and the ensuing price escalations for steel, concrete, and other major construction materials are risks that may seriously affect the ability to complete a project on time and within budget.

During the initial activities of the procurement process and contract development, the organization (buyer) will conduct out-reach using Request for Information or Requests for Interest to obtain feedback from potential contractors (seller). Buyer organizations usually rely on the expertise of the seller's project team to provide analysis and guidance on decisions relating to these issues and other enterprise environmental factors, including governmental and industry standards, stakeholder risk tolerances, and project management information systems.

.2 Organizational Process Assets

See Sections 12.1.1.2 and 4.1.1.4 of the *PMBOK® Guide*—Third Edition. The available organizational process assets among stakeholders for a construction project may vary widely. For example, a multi-national manufacturer may have lengthy and specific guidelines for the renovation of a plant; and yet a one-time construction purchaser may have no formalized procedures.

Organizations further down the contract hierarchy, such as subcontractors, also vary in the availability of process assets. A large subcontractor who performs work over a wide area may have well-developed procedures for purchasing from suppliers and sub-subcontractors. In contrast, a small subcontractor may rely on past knowledge without any specialized approaches.

.3 Project Scope Statement

See Section 5.2.3.1 of the *PMBOK® Guide*—Third Edition, which sets forth the various purposes of the project scope statement. These purposes include, among other things, descriptions of the project deliverables and the work required to create the deliverables, as well as providing a common understanding of the project scope among the stakeholders.

Large corporations that have significant and ongoing construction needs probably have established processes to develop a formal scope statement. Buyers that rarely utilize construction services can engage a representative, such as a designer or construction manager, to assist in the creation of a scope statement. sellers below the ultimate buyer, such as the general contractor and subcontractors, will generally utilize the documents created by the buyer for their own scope statements.

.4 Work Breakdown Structure

See Sections 12.1.1.4 and 5.3.3.2 of the *PMBOK® Guide*—Third Edition. The Work Breakdown Structure (WBS) is defined in as "a deliverable-oriented hierarchal decomposition of the work to be executed by the project team, to accomplish the project objectives and create the required deliverables." The development of the construction WBS considers labor jurisdictions, union agreements, contracting hierarchy, work phasing, and multiple-contract coordination.

All of the direct stakeholders in a construction project perform this process, to some extent. It is common for the ultimate buyer to designate other project participants to perform some of the tasks involved. For example, the Architect/Engineer firm (designer) prepares the contract documents as part of their work, which defines the total scope of the project. The construction manager, while planning the work with an eye towards scheduling and cost control, may subdivide the project tasks through the generation of work packages.

.5 WBS Dictionary

See also Sections 12.1.1.5 and 5.3.3.3 of the *PMBOK® Guide*—Third Edition. The WBS dictionary is the document that details the components of the WBS and may include such information as the identification of the responsible organization, a statement of work, and schedule milestones. Some project management software packages developed for the construction industry are capable of generating a form of WBS dictionary.

.6 Project Management Plan

See Section 12.1.1.6 of the *PMBOK® Guide*—Third Edition. The process of developing a project management plan, (Section 4.3 of the *PMBOK® Guide*—Third Edition), addresses the actions necessary to integrate and coordinate the various subsidiary plans, such as the quality, schedule, and risk management plans. For construction projects, many of the bases for the overall management plan, as well as the subsidiary plans, are set forth in the contract documents.

12.1.2 Plan Purchases and Acquisitions: Tools and Techniques

See Section 12.1.2 of the *PMBOK® Guide*—Third Edition, with modifications to Sections 12.1.2.1 and 12.1.2.3 and the addition of Section 12.1.2.4 as follows:

.1 Make-or-Buy Analysis

See also Section 12.1.2.1 of the *PMBOK® Guide*—Third Edition. In construction, this refers to self-performing of the work by the owner organization or the contracting for the work to be constructed by others.

.3 Contract Types

See also Section 12.1.2.3 of *PMBOK® Guide*—Third Edition. In construction, the contract type will vary over the project life cycle. Typically, reimbursable contracts are used for the conceptual and design work at the beginning of the project, and fixed-price contacts are preferred for the construction work. The following factors affect the selection of the contract type for a specific work package:

- Level of detail available.
- Urgency of the procurement.
- Level of competition desired.
- Level of competition available.
- Organization's risk utility or tolerance.

The major types of construction contracts are as follows:

- **Fixed-price or lump-sum contracts.** The contractor performs the work for a fixed, lump-sum price according to the contract bid package.
- **Unit-rate contracts.** The contractor performs the work for specified unit rates.
- **Cost-reimbursable contracts.** The contractor performs the work on a reimbursable cost basis plus a professional fee. The fee can be a fixed amount, a percentage of the costs up to a target amount, an incentive amount, or, other variations.
- **Time and materials contracts.** The contractor is reimbursed for the time spent and resources expended on the work performed.

.4 Project Delivery Methods

See also 12.1.1.4 of the *PMBOK® Guide*—Third Edition. In construction, a project delivery method refers to selecting the best strategy for project execution, including the project procurement activities. The strategic decisions made at this point determine which portions of the project are best suited for a particular execution approach depending upon a number of factors such as degree of definition, schedule, and cost requirements and uniqueness of requirement. Alternative project delivery methods include the following:

- **Design-bid-build (traditional method).** The design and the construction functions are performed by separate organizations. This is the traditional construction method where the construction is solicited by way of competitive bidding after the design is essentially complete.
- **Design-build.** The responsibility for both design and construction is obtained from a single source.
- **Turn-key.** The contractor has overall responsibility for delivering the project to an owner, providing all services for initial concept, design, purchasing, construction, commissioning, and start-up.
- **Construction management.** The construction management (CM) entity manages the overall functions of the project including design, bidding, purchasing, and construction. The CM can execute the construction on an agency or on an at-risk basis. During construction, the at-risk CM functions much like a general contractor with their subcontractors; whereas, the agency CM manages the work of prime contractors and their subcontractors.
- **Single-source, non-competitive.** In cases where the construction requirements are unique or where there is only one source for the desired result, a negotiated contract with the source is the usual way of accomplishing this.
- **Design-build-operate-maintain-transfer (DBOM).** This method encompasses the design-build method with the added feature of time scaled functions for operating and maintaining the product after construction is completed. After the DBOM contract is completed, the product is transferred to the buyer, and/or another contract is procured for continuation of the operating and maintaining functions.

12.1.3 Plan Purchases and Acquisitions: Outputs

See also Section 12.1.3 of the *PMBOK® Guide*—Third Edition, with the modification of 12.1.3.2 and the addition of 12.1.3.3 as follows:

.2 Contract Statement of Work

See also Section 12.1.3.2 of the *PMBOK® Guide*—Third Edition. In construction, the contract statement of work define the portions of the work to be performed by the particular construction entity; that is, the general contractor, prime constructor, or subcontractor. The contract statements of work define the outcomes (deliverables or products) of each work package, define the boundaries and interfaces with other work packages, set up the resource requirements anticipated, and specify the criteria by which to measure completion or satisfaction of the scope of work.

Most construction contracts have three major components: Introduction/Information for Bidders, General Provisions, and Technical, which includes specifications, drawings, and reference information. Contract format, content, and language are guided by best practices developed within the industry, including documented standards from organizations such as ISO, AIA, CMAA, and CSI.

.5 Long Lead Items

The design/engineering entity should identify the major equipment and other items to be purchased by the construction entity, which will take a considerable length of time to procure. The initiation of the procurement may be started by the design/engineering entity and then turned over to the construction entity for final purchase ordering and processing.

12.2 Plan Contracting

See Section 12.2 of the *PMBOK® Guide*—Third Edition, which states, "The plan contracting process prepares the documents needed to support the request seller responses process and select sellers' process." See also Request Seller Responses (Sections 12.3) and Select Seller (Section 12.4).

12.2.1 Plan Contracting: Inputs

See Section 12.2.1 of the *PMBOK® Guide*—Third Edition.

12.2.2 Plan Contracting: Tools and Techniques

See Section 12.2.2 of the *PMBOK® Guide*—Third Edition, with the modification of Section 12.2.2.1 as follows:

.1 Standard Forms

Refer to Section 12.2.2.1 of the *PMBOK® Guide*—Third Edition. Construction trade organizations, professional societies, and large owner organizations generate standard forms of contracts for use in contract development. These standard forms help reduce the time and expense for contract drafting and tender solicitation, standardize processes from project to project, and help ensure the quality of the final agreement. Sample organizations that make standard formats available include the Associated General Contractors (AGC), the American Institute of

Architects (AIA), and the Construction Owners Association of America (COAA). At the international level, the International Federation of Consulting Engineers (FIDIC) has developed a series of contractual standard forms that is being used globally.

12.2.3 Plan Contracting: Outputs

See Section 12.2.3 of the *PMBOK® Guide*—Third Edition, with modifications to Sections 12.2.3.1 through 12.2.3.3 as follows:

.1 Procurement Documents

See Section 12.2.3.1 of the *PMBOK® Guide*—Third Edition. The procurement documents should describe the tendering procedures and requirements which are to be used. The procurement documents will indicate the procurement process, the evaluation criteria and the requirements for information to be submitted by the sellers in the response. The contract requirements should:

- Specify items the buyer is expecting with the seller's response. The items, which are specified have requirements and are deliverables, can range from price data only to a larger list of information such as drawings, product data, preliminary bill of materials (BOM), company brochures/contract history, and qualifications of key personnel.
- Specify the process that the buyer will use to evaluate seller's response. The buyer establishes a brief narrative on how the bid information in the seller's response will be evaluated to determine the contract award. For Request for Quotations (RFQ), the price and bid information of the apparent low bid response is evaluated. For Request for Proposals (RFP), the bid information for all seller responses are evaluated and the criteria is usually listed in order of importance with the highest weighted criteria listed first.

.2 Evaluation Criteria

See Section 12.2.3.2 of the *PMBOK® Guide*—Third Edition, which states that "Evaluation criteria are used to rate or score proposals. They can be objective (e.g., 'the proposed project manager must be a certified Project Management Professional, PMP®') or subjective (e.g., 'the proposed project manager must have documented previous experience with similar projects'). See also Evaluation Criteria (Section 12.4.1.3 of this extension).

Evaluation criteria are often included as part of the procurement documents." The evaluation of the potential construction bidders (i.e., potential sellers) may occur in a two-stage process:

- **Pre-qualification.** Screening potential sources to establish a short list of bidders who possess the required technical and commercial capability to perform the work package.
- **Proposal evaluation.** Evaluating the technical and commercial proposal for the specific project.

This two-stage evaluation process saves time and effort for both the bidders and the proposal review process. The first review phase establishes if the interested bidders meet the minimum technical requirements. This involves evaluating the capabilities of potential bidders, including:

- Understanding of project requirements.
- Technical capability necessary to perform the work.

- Management approach, processes, and procedures to ensure a successful project.
- Financial capacity.

The second phase involves the technical and commercial evaluation of the tenders. In comparing several bids, the contract price of a particular bidder may be adjusted to reflect the effect of other cost factors, such as delivery time, price alternatives, or, costs associated with administering the contract. The final decision is based largely on the overall projected cost to the owner to perform the specific work packages, using the construction contractors under consideration. See also Section 12.4.1.3 on Evaluation Criteria of this extension.

.3 Contract Statement of Work

See Section 12.1.3.2 of the *PMBOK® Guide*—Third Edition. The statement of work (SOW) describes the facilities to be constructed in sufficient detail to allow potential bidders of the work package to determine if they are capable of providing the required construction services.

The SOW will address the nature of the facilities, the needs of the owner, and the expected contract form. In addition, the SOW will describe special requirements, if any, including collateral services, performance reporting, post-project operational support, and/or specific content and format requirements.

Typically, the architectural/engineering firm engaged in designing the project will develop the SOW. In cases where a turn-key or design-build contract type is used, the owner will provide the functional and aesthetic requirements of the facilities they envision in a Statement of Requirements (SOR). In addition, the SOW may solicit potential bidder input to propose solutions for certain problems. The SOW and associated tender documents should be clear, concise, and specific about the contract requirements.

12.3 Request Seller Responses

See Section 12.3 of the *PMBOK® Guide*—Third Edition.

12.3.1 Request Seller Responses: Inputs

See Section 12.3 of the *PMBOK® Guide*—Third Edition.

12.3.2 Request Seller Responses: Tools and Techniques

See Section 12.3.2 of the *PMBOK® Guide*—Third Edition, with modifications to Sections 12.3.2.1 through 12.3.2.3 and the addition of Section 12.3.2.4 as follows:

.1 Bidder Conferences

See also Section 12.3.2.1 of the *PMBOK® Guide*—Third Edition. In construction, bidder conferences are commonly referred to as pre-bid conferences, and often include a tour of the proposed site. Some buyers require participation by all major suppliers. In some non-public sector instances, bid requirements stipulate non-attendance at the bid conference precludes submission of a bid response.

.2 **Advertising**

See also Section 12.3.2.2 of the *PMBOK® Guide*—Third Edition. In addition to traditional classified advertising methods in print media, the Internet solicitation of potential bidders (sellers) is also growing in use.

.3 **Develop Qualified Sellers List**

See also Section 12.3.2.3 of the *PMBOK® Guide*—Third Edition. In construction, the owner (buyer) often relies on the expertise of the designer for a list of qualified bidders (sellers). The buyer's representative or designer will provide a list of potential bidders that meet the contract requirements and are recommended for solicitation of a bid/proposal. The criteria for assessing bidder eligibility for listing may include:

- Relevant construction experience. Addresses the experience of the potential bidder (seller) with similar projects;
- Identification of key members. Provides resumes, including descriptions of relevant work experience, of upper-level team members, such as project managers;
- Available resources. Description and availability of the proposed project resource elements—manpower, equipment, and materials;
- Quality programs. Description of and experience with quality programs, including quality planning, quality assurance and quality control;
- Safety. Description of and experience with safety programs. A verification of the bidder (seller) experience rate from its worker's compensation insurance carrier may be required;
- Surety bond. Demonstrates the ability to secure construction surety bonds for the appropriate amount and coverage;
- Insurance. Provides certificate of insurance for the coverage set forth in the contract documents;
- Previous disputes. Description of any claims of material breach of contract that lead to arbitration, litigation or some other form of dispute resolution proceedings;
- Regulatory requirements. Demonstrates the ability to comply with any special regulations for the project.

.4 **The Internet**

Certain buyers are now using Internet technology to solicit bids and real-time quotations from potential sellers for their projects. Commonly referred to as "reverse auction bidding," it is a controversial practice in competitive bidding scenarios, but has been included because of its increased usage and application to construction.

12.3.3 Request Seller Responses: Outputs

Refer to Section 12.3.3 of the *PMBOK® Guide*—Third Edition.

12.4 Select Sellers

See 12.4 of the *PMBOK® Guide*—Third Edition, which states "The Select Sellers process receives bids or proposals and applies evaluation criteria as applicable to select one or more sellers who are both qualified and acceptable as a seller."

The priced proposals of the contract bidders (sellers) should be compared against an independent estimate prepared by the owner's representative, to help analyze apparent discounts and premiums offered by the contract bidders. The evaluation of priced proposals against the independent estimate is to help ensure that the particular bidder can realistically perform the contract work at their stated price. A proposal offering an apparent discount may be intentional underbidding by a potential seller to "buy the job." Analyzing higher-priced bid proposals helps determine if the apparent premium is worth the cost. The award of the contract to a contractor that cannot properly perform the work for their proposed price will very likely cause grave difficulties during the contract execution, and, potentially a failed project.

There are situations where procurement of construction work may need to come from a single source. Depending on the organization's (buyer's) policies, the level of justification to preclude competitive bidding can vary. The most comprehensive justifications are required for public sector contracts where significant limitations are imposed by the organizations. Single sourcing may be considered in the following cases:

- A single bid/proposal response was submitted
- Time constraints during emergency conditions prevent use of the procurement cycle
- Technical uniqueness of requirements prevents non-original equipment manufacturer (OEM) from bidding
- Extended warranty on the product prevents the use of non-OEM contractors

12.4.1 Select Sellers: Inputs

See Section 12.4.1 of the *PMBOK® Guide*—Third Edition, with the modification of Section 12.4.1.3. In the traditional design-bid-build project delivery scenario, the proposal from a general/prime contractor will be an all-inclusive lump-sum bid price for the entire construction work. The general/prime contractor may perform all or part of the construction work, with the assistance of specialty subcontractors. The extent to which a general/prime contractor will subcontract work depends greatly on the nature of the project and the contractor's own organization. Customarily, the prime contractor will perform the basic project operations and will subcontract the remainder to various specialty contractors.

.3 Evaluation Criteria

Refer to Section 12.4.1.3 of the *PMBOK® Guide*—Third Edition. In construction, the evaluation of the seller's bids and proposals is dependent on the procurement method established by the buyer and reflected in the contract documents. The two primary methods are Request for Quotation (RFQ) and Request for Proposal (RFP). The RFQ process is a direct evaluation that is focused on the apparent low bidder. Provided the seller's submission contains all the items required by contract, the buyer evaluates the contractor's ability to provide the product. This evaluation includes reference checks on technical, management, quality, and safety aspects of prior contract performance as well as a review of financial viability of the company. Assuming there is no item that the contractor can not address, the award can be processed. The RFP process is normally a two-step process involving an evaluation of the technical proposal followed by a review and negotiation of the cost proposal.

12.4.2 Select Sellers: Tools and Techniques

See Section 12.4.2 of the *PMBOK® Guide*—Third Edition, with the addition of Section 12.4.2.8 as follows:

.8 Financial Modeling

Financial modeling may be used to assess the bidders' proposals, including the potential life cycle costs, and, as a means to control any potential bias of the selection process.

12.4.3 Select Sellers: Outputs

See Section 12.4.3 of the *PMBOK® Guide*—Third Edition.

12.5 Contract Administration

See Section 12.5 of the *PMBOK® Guide*—Third Edition, which states that Contract Administration is a two-way process which ensures that both the buyer and seller adhere to the requirements in their contract. Construction projects are multi-layered hierarchies and have numerous buyers and sellers, with many of the project stakeholders serving in both capacities. This includes the contractual relationships between the owner's representative(s) and the general/ prime construction contractor, and, the general/prime contractor and their subcontractors and suppliers.

12.5.1 Contract Administration: Inputs

See Section 12.5.1 of the *PMBOK® Guide*—Third Edition, with the following modifications:

.1 Contract

The contract is described in Section 12.4.3.2 of the *PMBOK® Guide*—Third Edition and this extension.

.2 Contract Management Plan

The contract management plan is described in Section 12.4.3.3 of the *PMBOK® Guide*—Third Edition and this extension.

.3 Selected Sellers

"Selected sellers" is described in Section 12.4.3.1 of the *PMBOK® Guide*—Third Edition and this extension.

.4 Performance Reports

Seller performance is documented in progress reports and payment requested is submitted to the buyer. The seller reports on a variety of topics specified in the contract and it includes work performed by its subcontractors, vendors, suppliers and testing facilities. As may be required by the buyer, independent field reports are submitted by on-site personnel such as the designer or construction manager.

.5 Approved Change Requests

Approved change requests, often referred to in construction terminology as change orders, reflect agreed-upon modifications to the contract scope, price, and/or schedule. Although most contract documents require changes to be in writing, the time sensitivity of construction projects often necessitates the recognition, approval, and processing of verbal change orders. The verbal change orders will be acknowledged between the buyer and seller, and later confirmed through a formal written change to the contract.

.6 Work Performance Information

Work performance information is another task that is often executed by the designer or other buyer's representative. For example, a general contractor's invoice for work performed in a given pay period is, in effect, a document providing performance information. The project designer or other buyer's representative will review the invoice for accuracy, thereby verifying the work was completed. The verbal change orders, when authorized by the contract, will be acknowledged between the buyer and seller, and later confirmed through a formal written change to the contract See also Performance Reporting (Section 12.5.2.4 of this extension).

12.5.2 Contract Administration: Tools and Techniques

See Section 12.5.2 of the *PMBOK® Guide*—Third Edition, with the following modifications:

.1 Contract Change Control System

The contract for construction, as discussed in Section 12.4.3.2 of the *PMBOK® Guide*—Third Edition and this extension, usually sets forth the procedures for the change control system to be utilized for a particular project. Standard provisions often include certain notice requirements, submittal procedures, and review responsibilities of the appropriate stakeholders.

.2 Buyer-Conducted Performance Review

As previously discussed, it is not unusual for the buyer to direct the designer or a consultant to conduct performance reviews of the seller's work. The review may address many issues, including adherence to the project schedule, the quality of the work in place, and budget issues. The designer, who is most familiar with the buyer's needs as addressed in the contract documents, is therefore in the best position to perform such a review.

.3 Inspections and Audits

According to Section 12.5.2.3 of the *PMBOK® Guide*—Third Edition, inspections and audits as directed by the buyer might cover processes or the deliverables themselves. In construction, such inspections usually focus on compliance with material and installation standards as set forth in the contract documents. An example is a core test of concrete to verify that the correct water-to-cement ratio, pursuant to the approved mix design, is present in the concrete slab.

.4 Performance Reporting

According to Section 12.5.2.4 of the *PMBOK® Guide*—Third Edition, "Performance reporting provides management with information about how effectively the seller is achieving the contractual objectives." The buyer will analyze the reports to ascertain the actual progress and adherence to other contract elements such as schedule, technical compliance and quality. The buyer will conduct meetings with the seller and field representatives to discuss, clarify, and determine performance against established criteria. The criteria may include physical progress, earned value, material delivered, and manpower usage.

.5 Payment System
Payment procedures for a construction project, as required by the buyer, are usually set forth in the contract. Each seller has its own system that must be sufficiently flexible to meet the buyer's requirements as to the form and timing of invoices, breakdown of costs, submittal of backup documentation, etc.

.6 Claims Administration
It is likely that there will be disputed or contested changes because of the substantial amount of information contained in construction documents. Thus, claims administration is an important component of all construction projects. Although the subject is referenced in Section 12.5.2.6 of the *PMBOK® Guide*—Third Edition, this extension devotes an entire chapter on Project Claim Management, for addressing claims (see Chapter 16).

.7 Records Management System
Record management systems are as varied as the projects they serve. Such systems help project managers to manage contract documents, and are used to assist in the retrieval and archiving of documents.

.8 Information Technology
Although construction is an industry that is known to move slowly towards change and even slower to embrace new technologies, project managers are now making use of information technology. For example, many projects now utilize project-specific websites for the dissemination of meeting minutes and scheduling information. Additionally, submittals for requests for information and the subsequent clarifications and revisions of the contract documents are now disseminated electronically.

12.5.3 Contract Administration: Outputs

See Section 12.5.3 of the *PMBOK® Guide*—Third Edition, with the following modifications to Sections 12.5.3.1, 12.5.3.2, 12.5.3.4, and 12.5.3.5 as follows:

.1 Contract Documentation
Construction projects generate more contract documentation than nearly any other kind of enterprise. Indeed, most construction contract documents require memorializing numerous aspects of the work, including schedules, change requests, shop drawings, etc. Other documentation that is not necessarily contractually mandated, but is often utilized in construction to support the work, includes daily logs, requests for information, progress photos, etc.

.2 Requested Changes
It is unlikely that any construction project will conclude without requiring changes to the project management or procurement plans. Integrated Change

Control (Section 4.6 of the *PMBOK® Guide*—Third Edition) lists several change management activities, including identifying a change, reviewing and approving a change, and controlling and updating affected requirements, such as scope, cost, and quality. See also Project Claims Management (Chapter 16 of this extension), which addresses disputed change requests.

.4 Organizational Process Assets (Updates)

Refer to Section 4.1.1.4 of the *PMBOK® Guide*—Third Edition. The available organizational process assets among stakeholders to a construction project could vary widely.

.5 Project Management Plan (Updates)

Updating a management plan usually will require updating of both the procurement plan and contract management plan. Any approved change orders to the contract documents, including revised drawings and specifications, will require the responsible project management team to make the appropriate updates, revisions, and amendments to all affected plans.

12.6 Contract Closure

See Section 12.6 of The *PMBOK® Guide*—Third Edition.

12.6.1 Contract Closure: Inputs

See Section 12.6.1 of the *PMBOK® Guide*—Third Edition, with the modification of Section 12.6.1.3 as follows:

.3 Contract Documentation

Refer to Section 12.5.3.1 of the *PMBOK® Guide*—Third Edition. The contract documentation includes the seller's submission of warranties. The warranties can encompass the seller's workmanship as well as material vendors' and manufacturers' warranties that meet or in many instances exceed the seller's warranty period.

For many projects, the document that initiates contract closure is a written communication, such as a letter or specified form, submitted by the constructor (seller) advising the designer and owner (buyer) that the seller has achieved substantial completion. The designer must then certify that substantial completion has been reached, or advise the seller as to why it has not. Closure documentation prepared and submitted by the constructor may include:

- Material warranties and workmanship guarantees.
- Equipment manufacturer warranties.
- Final inspection approvals from government authorities.
- Equipment manufacturer operation and maintenance manuals.
- As-built drawings.
- Sign-off sheets for training of owner personnel.

12.6.2 Contract Closure: Tools and Techniques

See Sections 12.6.2 of the *PMBOK® Guide*—Third Edition, with the addition of Section 12.6.2.3 as follows:

.3 Final Inspection and Acceptance Procedures

This tool is described in Section 5.4.3 of the *PMBOK® Guide*—Third Edition. When a construction contract is completed, a list of the remaining items (called a "punch list") is prepared, which documents all outstanding work to be performed by the constructor. When all items on the punch list are completed, the general contractor (constructor) requests a final inspection. The record of a final inspection testifying that all contract work is complete is mandatory for a proper contract closeout.

12.6.3 Contract Closure: Outputs

See Section 12.6.3 of the *PMBOK® Guide*—Third Edition.

Chapter 13

Project Safety Management

Project Safety Management processes include all activities of the project sponsor/owner and the performing organization which determine safety policies, objectives, and, responsibilities so the project is planned and executed in a manner that prevents accidents, which cause, or have the potential to cause, personal injury, fatalities, or property damage. For convenience, the term safety management is used throughout this extension to include both safety management and health management. Project Safety Management interacts with all other project management processes and process groups. See Figure 13-1 for an overview of the processes and respective inputs, tools and techniques, and outputs for Project Safety Management.

Active communication must be implemented to provide clarification to all stakeholders regarding the project's safety objectives and the implications of their execution. On urban projects, the neighbor community is a major stakeholder more so than in any other project, and special attention must be dedicated to the community's needs and expectations, especially regarding safety, which can have a great impact on the project, sometimes regardless of the existence of permits. However, because the issue of communication management is addressed in Chapter 10 of the *PMBOK® Guide*—Third Edition and this extension, it will not be detailed in this chapter.

Other major stakeholders include statutory authorities, usually comprised of representative bodies from local, regional and federal government, and, as in the case of nuclear and oil and gas, for example, industry regulators. These bodies have their own stakeholders and respond to them in the ways described in the *Government Extension to the PMBOK Guide Third Edition.*

The performing organization implements the safety management system through the policy, procedures, and processes of safety planning, safety assurance, and safety control, and undertaking continuous improvement activities throughout the project, as appropriate. As with quality and environmental management, safety management requires ensuring that the project management system employs all processes needed to meet project requirements, and that these processes take due cognizance of safety. Project Safety Management shares many common characteristics with Project Quality Management and Project Environmental Management, and it is for this reason, their requirements appear very similar. Project Safety Management, therefore, consists primarily of ensuring that the conditions of the contract (including those contained in legislation and any project technical safety specifications), are carried out to assure the safety of both those working on site and in the vicinity of the project. Project Safety Management must address both the management of the project and the product of the project (and its component parts), including assessing and determining how the different project management processes interact to fulfill the needs of the project, and whether changes or improvements are needed to accomplish the safety objectives of the project. Many believe that proper and effective project management would be incomplete without due consideration of the requirements for safety management. Furthermore, Project Safety Management needs to be integrated with Risk Management processes (See Chapter 11 of this extension) in order to accomplish the stated objectives.

Project safety management applies to all aspects of project management. As in the case of quality and environmental management, this broad application of safety management results in addressing three distinctive (and sometimes conflicting) sets of requirements, namely:

(a) **Mandatory statutory safety requirements** imposed by legislation, and enforced by statutory third party authorities in the region (geographical or otherwise) where the project is to be constructed. These are generally applicable to all construction projects regardless of application areas. Special statutory safety requirements are often imposed on projects in industries such as nuclear, power generation, oil and gas, railways, underground/mining, etc.

(b) **Customer safety requirements contained in conditions of contract** (defining how they require specific safety requirements to be undertaken and administered, and, the technical safety performance and acceptance criteria (defined in legislation, statutory instruments, and project specifications, which specify the technical safety requirements). Technical safety requirements frequently reference mandatory legislative requirements and can also incorporate those for quality management (see Chapter 8) and environmental management (see Chapter 14). Other requirements include those arising from enterprise environmental factors (see Section 4.1.1.3 of the *PMBOK® Guide*—Third Edition).

(c) **Requirements of the performing organization**, to satisfy the commercial needs (optimize profit, return on investment, etc.) and increase reputation in the market place, etc. Other requirements include those arising from organizational process assets (see Section 4.1.1.4) of the *PMBOK® Guide*—Third Edition).

These processes interact with each other and with processes of other Knowledge Areas as well. Each process can involve effort from one or more persons or groups of persons, based on the needs and complexity of the project. Each process occurs at least once in every project and occurs in one or more project phases, if the project is divided into phases. Although the processes are presented here as discrete elements with well-defined interfaces, in practice they may overlap and interact in ways not detailed here. Process interactions are discussed in Chapter 3 of the *PMBOK® Guide*—Third Edition.

The requirements of the Safety Planning, Perform Safety Assurance, and Perform Safety Control processes and activities which are detailed in this chapter are those generally considered applicable to construction projects most of the time. However, it is common for project sponsors or owners to invoke additional requirements, such as the following:

- Constraints local to the geographic region and application area where the project is destined (may depend on the scale, scope and complexity of the project);
- Specifying safety management systems standards, for example where general safety measures are considered insufficient to provide the assurance and control required;

- Industry-specific codes and standards, which define specific project safety performance and acceptance criteria.

It should be noted that the absence of a specific safety management program or system does not necessarily mean that the system employed by the performing organization is ineffective. Likewise, having a safety management system or program does not mean the performing organization will produce safety-compliant products or work.

See Figure 13-1 for an overview of the processes and respective inputs, tools and techniques, and outputs.

13.1 Safety Planning

Safety planning involves:

- Determining how to approach, plan and execute the requirements for project safety management.
- Determining the applicable requirements (safety standards, regulations, specifications, etc.), which define the criteria that will be employed to determine both the

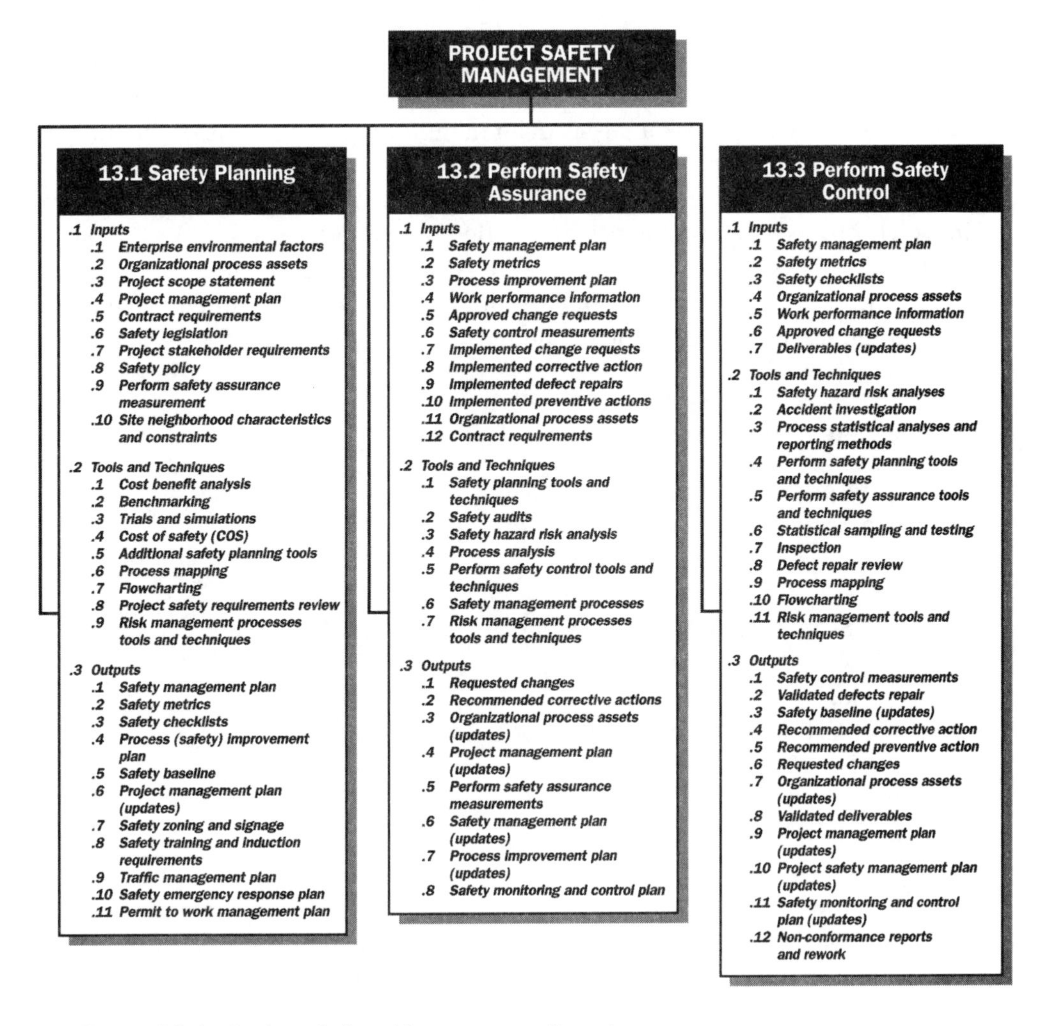

Figure 13-1. Project Safety Management Overview

suitability of a project management system in fulfilling the requirements of the project, and the ultimate acceptance of the product or the project.
- Assessing how best to apply the applicable acceptance criteria, documenting their characteristics and associated risks, and determining how to satisfy them.
- Determining how the suitability and effectiveness of project safety management will be assessed or demonstrated.

Safety standards incorporate project product codes, regulations, and standards, and they also include any condition of contract for which tangible deliverables have been defined and which will be used to determine safety acceptance and compliance. In other words, the contract *is* the principle project safety standard, as it will specify the applicable safety statutory and legislative requirements, technical safety codes, standards, and regulations.

Safety planning is one of the key processes when using the Planning Process Group (see Section 3.2.12 of the *PMBOK® Guide*—Third Edition) for the development of the project management plan (Section 4.3 of the *PMBOK® Guide*—Third Edition and this extension), and should be performed in parallel and would be an integral part of other project planning processes. Critical to the development of an effective project safety management system is an assessment as to how the performing organization should approach safety management, and also plan and execute the project to meet the requirements for safety. For example, the required changes in a product to meet identified or agreed-upon requirements (safety standards, regulations, specifications, etc.) may require cost or schedule adjustments; or the desired level of product safety may require a detailed risk analysis of an identified problem. One of the fundamental views of modern safety management is that safety must be planned, and built into processes, that is, safety requirements must be identified and built into the manner in which processes and activities are carried out as follows:
- Staged reviews for design safety before issued for construction.
- Ensuring safety requirements of hazardous products are identified and made known to potential users, for example, chemicals.

13.1.1 Safety Planning: Inputs

.1 Enterprise Environmental Factors
See Section 4.1.1.3 of the *PMBOK® Guide*—Third Edition.

.2 Organizational Process Assets
See Section 4.1.1.4 of the *PMBOK® Guide*—Third Edition.

.3 Project Scope Statement
See Section 5.2.3.1 of the *PMBOK® Guide*—Third Edition.

.4 Project Management Plan
See Section 4.3 of the *PMBOK® Guide*—Third Edition.

.5 Contract Requirements
See also Section 8.1.1.5 of this extension. Specifications, regulations, legislation, and standards (technical or legislative) are contractual requirements specific to construction projects.

Some construction projects may have additional requirements due to their nature, complexity, or specific industry application area. For example, there are mandatory application area-specific safety standards for construction within nuclear projects, oil and gas onshore and offshore projects, airport projects, or military projects, etc. In the construction industry, these requirements, issued by the project sponsor or owner, include a project scope statement, a description of the product(s) of the project, and references to all applicable standards and regulations.

.6 Safety Legislation

Safety legislation includes those requirements imposed by local, state or federal government, and/or application area regulators.

.7 Project Stakeholder Requirements

See Section 2.2 and Chapters 8 and 10 of the *PMBOK® Guide*—Third Edition.

.8 Safety Policy

The safety policy, although similar to the quality policy described in Section 8.1.1.1 of the *PMBOK® Guide*—Third Edition, differs in that it dictates much of the way in which construction activities are carried out from a safety perspective. The safety management policy also includes the degree to which the performing organization's management is committed to safety and can have a major impact on the effectiveness of a safety program.

.9 Perform Safety Assurance Measurements

Perform safety assurance measurements is also an input into Perform Safety Control (Section 13.3.1 of this extension) as an indicator of areas which may need further investigation. Perform safety assurance measurements is the result of perform safety assurance activities that are fed back into the Safety Planning Process (see Section 13.1.1) for use in re-evaluating and analyzing the performance of the planning activities of the performing organization and the standards and processes employed. When improvements are required, perform safety assurance measurements is used as an input into Perform Safety Control (see Section 13.3.1 of this extension) as an indicator of areas, which may need further investigation and to reassess the risks or decisions taken in early project phases.

.10 Site Neighborhood Safety Characteristics and Constraints

The characteristics of a construction site and its surrounding environment must be identified prior to project execution. For construction projects, the environment is the surrounding neighborhood where the project is to be undertaken, for which there may be constraints pertaining to safety management, quality management, and environmental management. These can include the proximity of adjacent residents, configuration of project offices, layout and location of construction equipment workshop, material delivery time constraints, traffic congestion in vicinity of the project site during peak periods, site security and access protocols, etc.

13.1.2 Safety Planning: Tools and Techniques

.1 Cost Benefit Analyses

See Section 8.1.2.1 of the *PMBOK® Guide*—Third Edition.

.2 **Benchmarking**

See Section 8.1.2.2 of the *PMBOK® Guide*—Third Edition.

.3 **Trials and Simulations**

Trials and simulations are discussed in Section 8.1.2.3 on Design of Experiments of the *PMBOK® Guide*—Third Edition. Examples include simulations of emergency responses procedures to assure that the controls developed are adequate to address those incidents identified as requiring an emergency response, and are very much dependent on industry application area constraints and requirements, i.e. mining, oil and gas, etc.

.4 **Cost of Safety (COS)**

Cost of safety (COS) is a similar process to the cost of quality described in Section 8.1.2.4 of the *PMBOK® Guide*—Third Edition. Determination of the cost of safety (COS) would usually include some form of cost-benefit analysis (see Section 8.1.2.1), an important distinction being that safety is of paramount importance regardless of cost.

.5 **Additional Safety Planning Tools**

Additional safety planning tools is similar to additional quality planning tools as described in Section 8.1.2.5 of the *PMBOK® Guide*—Third Edition.

.6 **Process Mapping**

Process mapping is commonly undertaken with flowcharting (see Section 13.1.2.7 of this extension) to:

- Map how a particular process is carried out;
- Determine how various processes interact;
- Identify any gaps in a particular work item or activity (termed gap-analysis), and include the absence of critical review points or a required deliverable (including the omission of verification that work has been undertaken and is acceptable. Such verification could relate not only to quality, but also safety or environmental issues).

.7 **Flowcharting**

See Section 8.3.2.3 of the *PMBOK® Guide*—Third Edition. A flowchart is a diagram that shows how various elements of a system or process (or series of processes) interact or interrelate. Flowcharting is commonly used with process mapping (see Section 13.1.2.6 of this extension) and with certain process statistical analyses and reporting methods. Flowcharting is also frequently used as a means of identifying non-value-added activities or functions, or delay points in tasks activities, and to define particular control points in work execution, for example, the issuance of a permit to enter prior to entering a confined space.

.8 **Project Safety Requirements Review**

Project requirements review includes an assessment and determination of the following:

- The characteristics and criteria of each activity and product(s) of the project, and how to satisfy them. These are sometimes incorporated into activity risk assessments.
- The applicable verification criteria, including those required to demonstrate acceptance and performance characteristics are fulfilled.

- Alternatives review and selection. In construction projects, it is common for some activities to be performed with different processes or arrangements for achieving the same result or output. This applies equally to safety management, quality management, and environmental management. Each process will require specific safety requirements. Examples include:
  - Rock formations can be removed by blasting or by using pneumatic breakers.
  - Effluents from chemical pipe cleaning can be treated at an on-site waste treatment works, or taken to an external treatment facility.
  - Inspections can be carried out by independent inspection organizations (either in-house or external), or undertaken by those carrying out the work (provided the competence of the latter to do so has been determined, assessed and agreed).

For each of these examples, the activity risk assessments may identify fundamentally different risks to workers.

Another example (and one that has seen significant increase in today's global economy), as with quality management, is when a requirement (standard or specification) that is developed in one geographical region is used in another region. In such instances, it is common for the characteristics and criteria of one region to differ in some degree from those in another location, as such requirements generally reflect the constraints of the region of origin. This is a case where compromise may be necessary, and requires the re-qualification of the requirements to ensure compliance. This compromise does not imply lowering standards for safety, but illustrates that the same end result can be achieved by different methods. This needs to be carefully scrutinized so as not to compromise safety and valid justification should be provided to project sponsors or owners, for obvious reasons.

Generally, all processes are analyzed to determine alternatives to increase effectiveness and efficiency. Examples are cost-benefit analyses (see Sections 8.1.2.1 of the *PMBOK® Guide*—Third Edition) and analyses in which time, cost, and safety considerations are balanced or even exceeded. Safety requirements can be mandatory constraints, as non-compliances can cause the project to have its execution permits cancelled, revoked, or otherwise not issued. Furthermore, a failure in any aspect of safety management can manifest itself in more significant failures in quality or environmental management.

- **Project stakeholder requirements review.** The requirements of project stakeholders are described in Section 2.2 of the *PMBOK® Guide*—Third Edition, and Section 8.1.2.8 of this extension, with the management of project stakeholder information described in Section 10.4 of the *PMBOK® Guide*—Third Edition.
- **Value engineering (VE).** Value engineering is a creative approach used on many construction projects to optimize product life cycle costs, save time, increase profits, improve safety, expand market share, solve problems, use resources more effectively, etc., and is frequently undertaken in conjunction with other perform quality planning tools and techniques such as cost-benefit analyses and design by experiments, etc.

.9 Risk Management Processes Tools and Techniques

Risk management processes tools and techniques are described in Chapter 11 of the *PMBOK® Guide*—Third Edition. They can be used to identify, qualify, and quantify safety risks, and support response planning. Risk management processes tools and techniques should be used in conjunction with all other tools and techniques described previously to ensure an integrated approach to quality, safety, and environmental management. The information to be used as input is available from the inputs of risk management planning, described in Section 11.1.1 of the *PMBOK® Guide*—Third Edition.

13.1.3 Safety Planning: Outputs

.1 Safety Management Plan

The project safety management plan essentially defines the strategy or methodology to be adopted by the performing organization to undertake safety management and how to fulfill the requirements of the project, and is therefore a subset of the overall project management strategy or methodology and, project management information system as described in Sections 4.1.2.2 and 4.1.2.3 of the *PMBOK® Guide*—Third Edition. The project safety management plan is a high-level strategic planning deliverable that defines the overall intentions and direction of the performing organization as expressed by top management, and is reviewed at various stages throughout the project to ensure the continued and sustained ability to cater for current and future phases of the project, and may include the participation of the project sponsor/owner, and other major project stakeholders, for example, industry regulators and local, national, and federal government. The safety management plan can include but is not limited to the following:

- **Safety Management Staffing or Human Resources Plan,** a subset of the project human resources plan developed as part of Section 9.1.3.3 of the *PMBOK® Guide*—Third Edition, developed by determining the various human resources arrangements, analyzing the risks and benefits of each, and selecting the optimum arrangement. The selected option must take into due consideration the nature of the work involved and necessary competencies required, contract and legislative requirements, responsibilities and accountabilities, organization structure, structures between the performing organization, project sponsor / owner and other project stakeholders and even the apportioning of work so as not to overload one particular function.
- **Project Safety Management Policy.** Described in Section 13.1.1.8 of this extension.
- **Safety Management Budget,** a subset of Project Cost Management described in Section 7 of the *PMBOK® Guide*—Third Edition. The safety management budget is developed by determining the different costs associated with the different approaches envisioned for the works to be carried out (including human resource arrangements), analyzing the costs associated with each, and determining the optimum costs and budgetary requirements for safety management.
- **Safety Records and Documentation Requirements.** The cornerstone of any safety management system is the records and associated documentation generated and employed not only as the basis to determine satisfactory or unsatisfactory performance, but also the effectiveness of the project management system as a whole. The requirement for records and documentation is a subset of the requirements described in Chapter 10 on Project Communications Management. Safety management records and supporting documentation are also employed when assessing compliance with the requirements of statutory safety requirement sand legislation.

- **Agreed Project Stakeholder Safety Requirements.** The agreed project stakeholder safety requirements form an input into Project Stakeholder Management described in Section 10.4 of the *PMBOK® Guide*—Third Edition, and include project stakeholders requirements for safety planning, safety assurance, and safety control.
- **Safety Reporting Requirements.** The reporting requirements for safety management are a subset of the overall project performance reporting requirements described in Section 10.3 of the *PMBOK® Guide*—Third Edition. Safety management reporting requirements would include, but are not limited to, assigned safety management staff and resources, safety baseline, safety management planned vs. actual expenditure, agreed performance and acceptance criteria, audit schedule, audits undertaken vs. those planned, details of audits (including periods for addressing non-satisfactory performance, details of corrective and preventive actions, statistical measurements to demonstrate project safety efficiency and effectiveness of the project management system, etc.
- **Project Safety Execution Constraints.** The project execution constraints imposed must be taken into account, and may require additional mandatory quality, safety and environmental requirements. Examples of such constraints would be rock blasting, working with noisy machines, removing vegetation / debushing, etc. At times, project execution characteristics may be dictated by the configuration of the construction site and that of the surrounding environment. Project execution constraints can often influence the approach or strategy adopted for quality, safety or environmental management and, therefore, form an input into Project Management (refer to Chapter 11 of the *PMBOK® Guide*—Third Edition).
- **Agreed Safety Performance and Acceptance Criteria.** Criteria for the safety performance and acceptance criteria for all attributes of the project must be finalized and agreed with the project sponsor or owner and, where applicable, any project stakeholder, and form the basis for the safety management baseline (see 13.1 .3.5 of this extension).
- **Project Administrative and Contract Closure Procedures.** Described in Sections 4.7.3.1 and 4.7.3.2 of the *PMBOK® Guide*—Third Edition. Administration and contract closure will include collation and assembly of all pertinent safety management records and other supporting documentation necessary to demonstrate that the safety requirements of the project have been fulfilled.
- **Operational Definitions.** The operational definitions describe in specific terms what something is and how it is to be measured. Safety assurance and control processes are frequently employed to determine how processes and project work are to be measured. For example, it is insufficient merely to state that management reviews of the project management system will be undertaken once per year or site safety inspections will be carried out as per frequencies specified in the contract or the hazard caused by construction equipment will remain within the permissible limits or all site staff will wear personal protective equipment. The project management team must, in relation to the above statements, be in a position to demonstrate tangibly that management reviews were undertaken or that the specific inspections were carried out or whether any oil spills or gas emissions were measured, and whether only heavy vehicles or all vehicles were inspected or that all site personnel were, indeed, issued with personal protective equipment.

- **Safety Communication.** (described in Section 10.3 of the *PMBOK® Guide*—Third Edition). This type of communication can cover a broad range of activities such as:
  - Barriers, signs, and bulletin boards (refer to Section 13.1.3.7).
  - Initial safety indoctrination meetings, tool box meetings, individual bulletins on a specific subject, etc. (See Section 13.1.3.8 of this extension).
  - Safety reports, including those required by legislation.

.2 Safety Metrics

Safety metrics are similar to quality matrices described in Section 8.1.3.2 of the *PMBOK® Guide*—Third Edition.

.3 Safety Checklists

Safety checklists are similar to the quality checklists described in Section 8.1.3.3 of the *PMBOK® Guide*—Third Edition.

.4 Process (Safety) Improvement Plan

See Section 8.1.3.4 of the *PMBOK® Guide*—Third Edition.

.5 Safety Baseline

Safety baseline is similar to quality baseline described in Section 8.1.3.5 of the *PMBOK® Guide*—Third Edition.

.6 Project Management Plan (Updates)

See Section 8.1.3.6 of the *PMBOK® Guide*—Third Edition.

.7 Safety Zoning and Signage

Generally, a project site can be divided into specific zones, each having its own particular safety requirements (e.g., workshops, storage areas, the different areas of the construction job-site, limited or controlled access areas, etc.). The zoning of a particular area of a project site assists in determining specific safety hazards or risks associated with each. Consequently, the appropriate signage forewarns construction personnel of particular hazards or particular controls to safeguard them. Such signage can include general signage to warn of access restrictions or more specific signage to warn of particular hazards. This process utilizes the principle of Visual Operations Management (VOM), to reinforce and repeat any specific instruction that would have been given during safety training and induction (see 13.1.3.8 in this extension). It is also common to employ safety signage banks at specific access points where all necessary safety signage relating to a specific area of the project is cited. These banks would be subject to audit or inspection at regular intervals to ensure that all appropriate signs are present and determine whether any have been damaged or otherwise lost, and need replacement.

.8 Safety Training and Induction Requirements

The requirements for safety training are generally divided into one of the following:

- **Safety training.** Safety training is common on all construction projects and includes mandatory statutory training (e.g., for operation of construction equipment) and undertaking of specific safety tasks (e.g., heavy-lifting operations). It has also become increasingly common that all construction personnel undergo mandatory general statutory safety training before permitted entrance onto a construction project site. This is a result of the increased occurrence of accidents which would, under normal circumstances, be seen as easily avoidable.

- **Safety inductions.** Safety inductions are also common on all construction projects and are generally task-specific, and relate to specific construction activities or tasks which address the particular risks associated with these tasks and the necessary mitigation measures to be employed.
- **Tool box meetings** Tool box meetings are generally undertaken to provide specific instructions regarding the use of certain construction equipment and tools, and include such content as, safety, operation, maintenance, etc.

Critical to any safety program is the need to plan when elements of the program need to be implemented to address specific parts of the project. Safety training planning is a subset of the human resources planning outputs (see 9.1.3 of the *PMBOK® Guide*—Third Edition).

.9 Traffic Management Plan

The traffic management plan defines the controls to be exercised over traffic in the vicinity of the project site including, but not limited to, project site entry and egress arrangements (including security checks), time limitations for deliveries, use of temporary roads for public traffic, weight restrictions, traffic signals and channeling of vehicles to avoid construction works, and access and egress for emergency vehicles, etc.

.10 Safety Emergency Response Plan

The necessity for a project-specific emergency response management plan is generally dictated by the constraints of the project, its environs, and industry application area, and is usually developed in conjunction with project sponsors and owners. For example, underground works (such as tunneling) have specific mandatory requirements relating to emergency response needs. Requirements may include, but are not limited to, responsibility and authority for key members of the emergency response team, communication requirements (especially of initial occurrence of emergency incidents, and with emergency service, local hospitals, etc.), provision of appropriate emergency response equipment, access and egress requirements for emergency response vehicles (fire, police, ambulance, etc.). Emergency responses should not be limited to safety incidents and can also include environmental emergency incidents, for example, the inadvertent discharge of contaminated material into water courses, which could lead to the contamination of reservoirs. In addition, it is common for application area regulators to require specific mandatory controls to be implemented as part of emergency response activities.

It is prudent to develop an integrated emergency response plan to address both safety and environmental incidents, especially as an emergency incident will directly or indirectly have safety and environmental implications.

.11 Permit to Work Management Plan

Many construction project application areas have mandatory requirements relating to permits, for example, permit to excavate, hot work permits, confined work permits, permit to enter bio-hazard areas, etc. These permit procedures are generally defined in the conditions of contract, and generally reflect the safety checks that need to be addressed prior to commencing specific activities.

Although the permit to work management plan is also a subset of the project communication management plan, the consequences of safety failures warrant its placement in this chapter. This is especially true where part of the scope of work for a construction project relates to the mitigation measures to deal with contamination originating from adjacent industries.

13.2 Perform Safety Assurance

Perform safety assurance involves:

- Applying the planned, systematic safety activities to ensure that the project employs all processes needed to meet requirements:
- Determining whether these processes (and their integration) are effective in ensuring that the project management system will fulfill the requirements of the project and the product of the project.
- Evaluating the results of safety management on a regular basis to provide confidence that the project will satisfy the relevant safety standards.

13.2.1 Perform Safety Assurance: Inputs

.1 Safety Management Plan

See Section 13.1.3.1 of this extension. The safety management plan describes how safety assurance will be applied and performed on the project, and defines the primary attributes of the project safety management plan which form inputs to the perform safety assurance process, for example:

- **Safety management human resources plan.** See Chapter 10 of the *PMBOK® Guide*—Third Edition.
- **Project safety management policy.** See Section 13.1.1.8 of this extension.
- **Safety management budget.** See Chapter 7 of the *PMBOK® Guide*—Third Edition.
- **Records and documentation requirements.** See also Chapter 10 on Project Communication Management of the *PMBOK® Guide*—Third Edition. Safety management records and supporting documentation are also employed when assessing compliance with the statutory safety requirements;
- **Agreed project stakeholder safety requirements.** See Section 10.4 of the *PMBOK® Guide*—Third Edition.
- **Reporting requirements.** See Section 10.3 of the *PMBOK® Guide*—Third Edition.
- **Project safety execution constraints.** See Sections 13.1.3.1 and 13.1.1.9 of this extension.
- **Agreed safety performance and acceptance criteria**. See Section 13.1.3.1 of this extension.
- **Project administrative and contract closure procedures.** See Sections 4.7.3.1 and 4.7.3.2 of the *PMBOK® Guide*—Third Edition.
- **Operational definitions.** See Section 13.1.3 of this extension.

.2 Safety Metrics

Safety metrics are similar to quality matrices described in Section 8.1.3.2 of the *PMBOK® Guide*—Third Edition.

.3 **Process Improvement Plan**
See Section 8.1.3.4 of the *PMBOK® Guide*—Third Edition.

.4 **Work Performance Information**
See Sections 4.4.3.7 and 10.3.3.1 of the *PMBOK® Guide*—Third Edition.

.5 **Approved Change Requests**
See Section 4.4.1.4 of the *PMBOK® Guide*—Third Edition.

.6 **Safety Control Measurements**
See Section 13.3.3.1 of the *PMBOK® Guide*—Third Edition.

.7 **Implemented Change Requests**
See Section 4.4.3.3 of the *PMBOK® Guide*—Third Edition.

.8 **Implemented Corrective Action**
See Section 4.4.3.4 of the *PMBOK® Guide*—Third Edition.

.9 **Implemented Defect Repairs**
See Section 4.4.3.6 of the *PMBOK® Guide*—Third Edition.

.10 **Implemented Preventive Actions**
See Section 4.4.3.5 of the *PMBOK® Guide*—Third Edition.

.11 **Organizational Process Assets**
See Section 4.1.1.4 of the *PMBOK® Guide*—Third Edition.

.12 **Contract Requirements**
See Section 13.1.1.5 of this extension.

13.2.2 Perform Safety Assurance: Tools and Techniques

.1 **Safety Planning Tools and Techniques**
See Section 13.1.2 of this extension. These techniques are similar to the quality planning tools and techniques described in Sections 8.1.2 of the *PMBOK® Guide*—Third Edition.

.2 **Safety Audits**
Audits are described in Section 8.2.2.2 of the *PMBOK® Guide*—Third Edition. Audits involve undertaking structured and independent review to ensure that project activities of the performing organization(s) comply with the project requirements and that such activities are suitable to fulfill the requirements of the project.

Safety audits of the product(s) of the project are called safety technical or safety compliance audits, for example Road Safety Audits, where an assessment is made of the measures implement for traffic management. These include an evaluation of results or outputs of activities vs. the performance and acceptance criteria defined in technical safety standards and specifications, to determine fitness for the purpose intended.

Safety audits can also be undertaken for the project management system as a whole or for individual component parts), for example, procurement management system (refer to Section 12.5.2.3 of the *PMBOK® Guide*—Third Edition), design management system, or commissioning management system, etc. Audits are also carried out to assess compliance with statutory and legislative safety requirements.

Integrated audits are commonly adopted (for example, incorporating the applicable requirements such as those for quality, safety, and environmental management) to provide a more accurate measure of the effectiveness of a specific area of work in fulfilling project requirements and are used to assess the effectiveness of the controls employed on a project as a whole rather than individually

.3 Safety Hazard Risk Analyses

Safety hazard risk analyses are systematic reviews of each construction process, activity, or work element to identify the potential safety hazards for project personnel as well as others who are present on the site associated with the activity or process. These analyses are normally carried out by specific and knowledgeable members of the project management team of the performing organization with the assistance of key construction supervisors and are part of the risk identification process outlined in Section 11.2 of the *PMBOK® Guide*—Third Edition.

.4 Process Analysis

Process analysis is described in Sections 8.2.2.3 of the *PMBOK® Guide*—Third Edition. Process analyses may also include statistical analyses (see 8.3.2.1 of the *PMBOK® Guide*—Third Edition).

.5 Perform Safety Control Tools and Techniques

Perform safety control tools and techniques is described as part of perform quality control tools and techniques in Sections 8.3.2 of the *PMBOK® Guide*—Third Edition, and Section 13.1.2 of this extension.

.6 Safety Management Reviews

Safety management reviews are discussed as part of management reviews detailed in Section 8.2.2.5 of this extension.

.7 Risk Management Processes Tools and Techniques

Risk management tools and techniques are described in Chapter 11 of the *PMBOK® Guide*—Third Edition.

13.2.3 Perform Safety Assurance: Outputs

.1 Requested Changes

See Sections 4.4.3.2 and 8.2.3.1 of the *PMBOK® Guide*—Third Edition.

.2 Recommended Corrective Actions

See Section 8.2.3.2 of the *PMBOK® Guide*—Third Edition.

.3 Organizational Process Assets (Updates)

See Section 8.2.3.3 of the *PMBOK® Guide*—Third Edition.

.4 **Project Management Plan (Updates)**
See Section 8.2.3.4 of the *PMBOK® Guide*—Third Edition.

.5 **Perform Safety Assurance Measurements**
Perform safety assurance measurements is a process discussed as part of perform quality assurance measurements in Section 13.1.1.9 of this extension. Perform safety assurance measurements is the result of perform safety assurance activities that are fed back into the Safety Planning Process (see Section 13.1.1) for use in re-evaluating and analyzing the performance of the performing organization, and the standards and processes employed. Perform safety assurance measurements is also used as input into Perform Safety Control (Section 13.3.1) as an indicator of areas which may need further investigation.

.6 **Safety Management Plan (Updates)**
See Section 13.1.3.1 of this extension.

.7 **Process Improvement Plan (Updates)**
See Section 8.1.3.4 of the *PMBOK® Guide*—Third Edition.

.8 **Safety Monitoring and Control Plan**
The safety monitoring and control plan describes how the project management team will implement the necessary safety controlling activities of the performing organization. The safety monitoring and control plan is a component or subsidiary plan of the project management plan (see Section 4.3 of the *PMBOK® Guide*—Third Edition).

The safety monitoring and control plan will either contain or make reference to specific procedures to be employed to ensure the safety compliance of the work that is carried out.

While the project safety management plan details how the performing organization will manage safety on the project, the project safety monitoring and control plan defines the actual monitoring and control activities to be employed and undertaken, especially:

- Items of work to be monitored;
- Reference to the applicable reference document and acceptance criteria;
- Applicable verification activities (inspection, tests, reviews, submissions, etc.) will be performed, and when such activities are performed in relation to the overall process;
- Responsibility for undertaking the work and each verification activity;
- Applicable characteristics and measurements to be taken or recorded;
- Applicable supporting documentation to be generated to demonstrate satisfactory or unsatisfactory performance.

To be effective, safety monitoring and control (and verification) must be integrated into how the physical work is performed. This process establishes control points or gates throughout the process to ensure that the next phase work will not proceed until the preceding work has been completed and verified as complete and compliant.

An emerging trend worldwide is the use of integrated execution and verification plans (EVPs) that incorporate (integrate) the necessary verification activities into actual work processes. EVPs also serve to plan work processes and the sequence by which such work needs to be performed. A similar approach can be adopted for monitoring and controlling activities for Quality Management (see Chapter 8) and Environmental Management (see Chapter 14), with EVPs containing not only the precise sequence of work to be performed but also all necessary monitoring and controlling verification activities.

13.3 Perform Safety Control

Perform safety control covers the following:

- Determining and applying measures for monitoring the achievement of specific project results throughout the project to determine whether they comply with the safety requirements.
- Identifying unsatisfactory performance and identifying ways to eliminate causes of unsatisfactory safety performance. This includes failures on the part of safety planning and safety assurance.

13.3.1 Perform Safety Control: Inputs

.1 Safety Management Plan

See Section 13.1.3.1. The safety management plan describes how safety management will be applied and performed on the project. Section 13.1.3.1 defines the primary attributes of the project safety management plan which form inputs to the perform safety control process, as follows:

- **Safety management human resources plan.** See Chapter 10 of the *PMBOK® Guide*—Third Edition. The organizational interface among the performing organization, the project sponsor/owner, and other applicable project stakeholders is important for effective project environmental control.
- **Project safety management policy.** See Section 13.1.1.8.
- **Safety management budget.** See Section 7 of the *PMBOK® Guide*—Third Edition.
- **Records and documentation requirements.** See Chapter 10 on Project Communications Management. Safety management records and supporting documentation are also employed when assessing compliance with the statutory safety requirements.
- **Agreed project stakeholder safety requirements.** See Section 10.4 of the *PMBOK® Guide*—Third Edition.
- **Reporting requirements.** See Section 10.3 of the *PMBOK® Guide*—Third Edition.
- **Project execution constraints.** See Sections 13.1.3.1 and 13.1.1.10.
- **Agreed safety performance and acceptance criteria**. See Section 13.1.3.1.
- **Project administrative and contract closure procedures.** See Sections 4.7.3.1 and 4.7.3.2 of the *PMBOK® Guide*—Third Edition.

.2 Safety Metrics

Safety metrics are similar to quality matrices as described in Section 8.1.3.2 of the *PMBOK® Guide*—Third Edition.

.3 Safety Checklists

Safety checklists are similar to quality checklists as described in Section 8.1.3.3 of the *PMBOK® Guide*—Third Edition.

.4 Organizational Process Assets
See Section 4.1.1.4 of the *PMBOK® Guide*—Third Edition.

.5 Work Performance Information
See Sections 4.4.3.7 and 10.3.3.1 of the *PMBOK® Guide*—Third Edition.

.6 Approved Change Requests
See Section 4.4.1.4 of the *PMBOK® Guide*—Third Edition.

.7 Deliverables (Updates)
See Section 4.4.3.1 of the *PMBOK® Guide*—Third Edition.

13.3.2 Perform Safety Control: Tools and Techniques

.1 Safety Hazard Risk Analyses
See Section 13.2.2.3 of this extension.

.2 Accident Investigation
It is important that each accident be investigated as to cause (direct and/or indirect), and a complete report should be made stating what happened and why. These reports are usually required by the insurance companies covering the resulting losses but the reports are also vital as a measure of, and for the improvement of, the performing organization's safety performance. In some cases, there reports are required by law enforcement agencies.

.3 Process Statistical Analyses and Reporting Methods
Process statistical analyses are described in Section 8.3.2 of the *PMBOK® Guide*—Third Edition. Common statistical analyses and reporting methods for safety management include:

- **Cause and effect analyses**. See Section 8.3.2.1 of the *PMBOK® Guide*—Third Edition.
- **Control charts**. See Section 8.3.2.2 of the *PMBOK® Guide*—Third Edition and Section 8.3.2.2 of this extension.
- **Histograms**. See Section 8.3.2.4 of the *PMBOK® Guide*—Third Edition.
- **Pareto analyses**. See Section 8.3.2.5 of the *PMBOK® Guide*—Third Edition and Section 8.3.2.5 of this extension.
- **Run analyses**. See Section 8.3.2.6 of the *PMBOK® Guide*—Third Edition.
- **Scatter analyses and diagrams**. See Section 8.3.2.7 of the *PMBOK® Guide*—Third Edition.

In addition, for construction, the applicable statistical methods of analyses and their respective reporting methods are generally defined in applicable standards and specifications for specific project products, as described in Section 8.2.1.6 of this extension.

.4 Perform Safety Planning Tools and Techniques
See Section 13.1.2 of this extension.

.5 Perform Safety Assurance Tools and Techniques
See Section 13.2.2 of this extension.

.6 Statistical Sampling and Testing
Statistical sampling is described in Section 8.3.2.8 of the *PMBOK® Guide*—Third Edition.

.7 Inspection
See Section 8.3.2.9 of the *PMBOK® Guide*—Third Edition.

.8 Defect Repair Review
See Section 8.3.2.10 of the *PMBOK® Guide*—Third Edition. Although defects repair would generally be considered an attribute of quality management (see Chapter 8), the review of defects with regard to any potential safety issues is common. Defects repair review can therefore commonly be addressed by means of the non-conformance control process.

.9 Process Mapping
See Section 13.1.2.6 of this extension.

.10 Flowcharting
See Section 8.3.2.3 of the *PMBOK® Guide*—Third Edition, and Section 13.1.2.7 of this extension.

.11 Risk Management Processes Tools and Techniques
Risk management processes tools and techniques are described in Chapter 11 of the *PMBOK® Guide*—Third Edition.

13.3.3 Perform Safety Control: Outputs

.1 Safety Control Measurements
Similar to quality control measures described in Section 8.3 of this extension, these measurements represent the results of safety control activities that are fed back as inputs into the Perform Safety Planning (see Section 13.1.1 of this extension) and Perform Safety Assurance (see Section 13.2.1 of this extension) processes of the performing organization to re-evaluate and analyze the performance of the project safety management system, including the application of contract requirements (see Section 13.1.1.5 of this extension), and safety legislation (see Section 13.1.1.6 of this extension).

.2 Validated Defects Repair
See Sections 4.4.1.6 and 8.3.3.2 of the *PMBOK® Guide*—Third Edition.

.3 Safety Baseline (Updates)
See Section 13.1.3.5 of the *PMBOK® Guide*—Third Edition.

.4 Recommended Corrective Action
See Section 4.5.3.1 of the *PMBOK® Guide*—Third Edition and Section 13.3 .3.4 of this extension.

.5 Recommended Preventive Action
See Section 4.5.3.2 of the *PMBOK® Guide*—Third Edition and Section 13.3 .3.5 of this extension.

.6 Requested Changes

See Section 4.4.3.2 of the *PMBOK® Guide*—Third Edition and Section 13.3.3.6 of this extension.

.7 Organizational Process Assets (Updates)

See Section 4.1.1.4 of the *PMBOK® Guide*—Third Edition.

.8 Validated Deliverables

See Section 13.3.3.7 of this extension.

.9 Project Management Plan (Updates)

See Section 4.6 of the *PMBOK® Guide*—Third Edition.

.10 Project Safety Management Plan (Updates)

See Section 13.1.3.1 of this extension.

.11 Safety Monitoring and Control Plan (Updates)

See Section 13.2.3.8 on this extension

.12 Non-Conformance Reports and Rework

See Section 8.3.3.13 of this extension.

Chapter 14

Project Environmental Management

Project environmental management processes include all the activities of the project sponsor/owner and the performing organization which determine environmental policies, objectives, and, responsibilities, the purpose of which is to minimize the impact on the surrounding environment and natural resources and to operate within the limits stated in legal permits. Project environmental management includes identifying the environmental characteristics surrounding a construction site and the potential impact which the construction may have on the environment; planning how to prevent environmental impacts; achieving environmental conservation and improvement, if possible; auditing the plan and controlling the results; and inspecting environmental conditions.

Active communication must be maintained with all stakeholders to provide clarification of the project's environmental objectives and the environmental implications of its execution. On urban projects, the neighboring community is a major stakeholder more so than in any other kind of project, and special attention must be paid to the community's environmental needs, expectations, and concerns, which can have a great impact on the project, regardless of the existence of permits. Communication management is addressed in detail in Chapter 10 of the *PMBOK® Guide*—Third Edition and in this extension, therefore it will not be detailed further in this section.

Other major stakeholders include statutory authorities, usually comprised of representative bodies from local, regional, and federal governments, as well as industry regulators, as in the case of nuclear and oil and gas projects, for example. These bodies have their own stakeholders and respond to them in the ways described in the *Government Extension to the PMBOK® Guide*—Third Edition.

The performing organization implements the environmental management system through the policy, procedures and processes of environmental planning, environmental assurance, and environmental control, and performing continuous improvement activities throughout the project, as appropriate. As with quality and safety management, environmental management requires ensuring that the project management system employs all processes needed to meet the project requirements, and that these processes take into consideration the environment. Project Environmental Management shares many common characteristics with Project Quality Management and Project Safety Management, and it is for this reason, their requirements appear very similar. Project Environmental Management therefore consists primarily of ensuring that the conditions of the contract (including those contained in legislation and any project technical environmental specifications), are carried out to minimize the impact the project will have on the environment, not only in the vicinity of the project, but also on environments far from the project site. Project Environmental Management must address both management of the project and the product of the project (and its component parts), including assessing and determining how the different project management processes interact to fulfill the needs of the project, and whether changes or improvements are needed to accomplish the environmental objectives of the project. Many believe that proper and effective project management would be incomplete without due consideration of the requirements for environmental management. Furthermore, Project Environmental Management must be integrated with Risk Management processes (See Chapter 11 of this extension) in order to accomplish the stated objectives.

The project management team must be careful not to confuse environmental management with the absence of environmental impact. Construction projects cause environmental impacts, due to their very nature. The goal of a good environmental management plan is to keep the impacts to an absolute minimum and within the limits foreseen in the legal permits.

Project environmental management applies to all aspects of project management, which entails addressing three distinct (and sometimes conflicting) sets of requirements, namely:

(a) **Mandatory statutory environmental requirements,** imposed by legislation and enforced by statutory third party authorities in the region (geographical or otherwise) where the project is to be constructed. These are generally applicable to all construction projects regardless of application areas. Special statutory environmental requirements are often imposed on project in industries such as nuclear, power generation, oil and gas, etc.

(b) **Customer environmental requirements contained in conditions of contract**, defining how they require specific safety requirements to be undertaken and administered, and, the technical safety performance and acceptance criteria (defined in legislation, statutory instruments, and project specifications, which specify the technical environmental requirements). Technical environmental requirements frequently reference mandatory legislative requirements and can also incorporate those for quality management (see Chapter 8 of this extension) and safety management (see Chapter 13 of this extension). Other requirements include those regarding enterprise environmental factors (see Section 4.1.1.3) of the *PMBOK® Guide*—Third Edition).

(c) **Requirements of the performing organization**, to satisfy the commercial needs (optimize profit, return on investment, etc.) and increase reputation in the market place, etc. Other requirements include those arising from organizational process assets (see Section 4.1.1.4) of the *PMBOK® Guide*—Third Edition).

The Project Environmental Management processes include the following:

14.1 Environmental Planning—Environmental planning involves the following:

- Determining how to approach, plan, and execute the requirements for project environmental management;
- Identifying the characteristics of the environment surrounding the construction site and determining which environmental standards are relevant to the project;

- Determining the impact that the project will have on the environment; and determining how to satisfy these standards.

Environmental standards do not only mean project codes, regulations, and standards; but they also include any condition of the contract for which tangible deliverables have been defined and which will be used to determine acceptance. In other words, the contract *is* the principle project environmental standard, as it will specify the applicable environmental statutory and legislative requirements and technical environmental codes, standards, and regulations.

14.2 Perform Environmental Assurance—Perform environmental assurance involves the following:

- Applying the planned, systematic environmental activities to ensure that the project employs all processes needed to meet environmental requirements;
- Determining whether these processes (and their integration) are effective in ensuring the project management system will fulfill the environmental requirements of the project and the product of the project;
- Evaluating the results of environmental management on a regular basis to provide confidence that the project will satisfy the relevant environmental standards.

14.3 Perform Environmental Control—Perform environmental control involves the following:

- Determining and applying measures for monitoring the achievement of specific project environmental results throughout the project to determine whether they comply with the requirements and identifying unsatisfactory environmental performance;
- Identifying ways to eliminate causes of unsatisfactory environmental performance. This includes identifying failures on the part of environmental planning and performing environmental assurance.

These processes interact with each other and with processes of other Knowledge Areas, as well. Each process can involve effort from one or more persons or groups of persons, depending on the need and complexity of the project. Each process occurs at least once in every project and occurs in one or more project phases, if the project is divided into phases. Although the processes are presented here as discrete elements with well-defined interfaces, in practice they may overlap and interact in ways not detailed here. Process interactions are discussed in Chapter 3 of the *PMBOK® Guide*—Third Edition.

The requirements of the Environmental Planning, Perform Environmental Assurance, and Perform Environmental Control activities detailed in this chapter are those generally considered applicable to construction projects most of the time. However, it is common for project sponsors or owners to invoke additional requirements, for example:

- Depending upon the scale, scope, and, complexity of the project, including constraints specific to the geographical region and application area where the project is destined, etc.;
- Specifying environmental management systems standards, for example, the ISO 14000 Series developed by such bodies as the International Organization for Standardization (ISO), where general environmental measures are considered insufficient to provide the assurance and control required;
- Industry-specific codes and standards, which define specific project environmental performance and acceptance criteria.

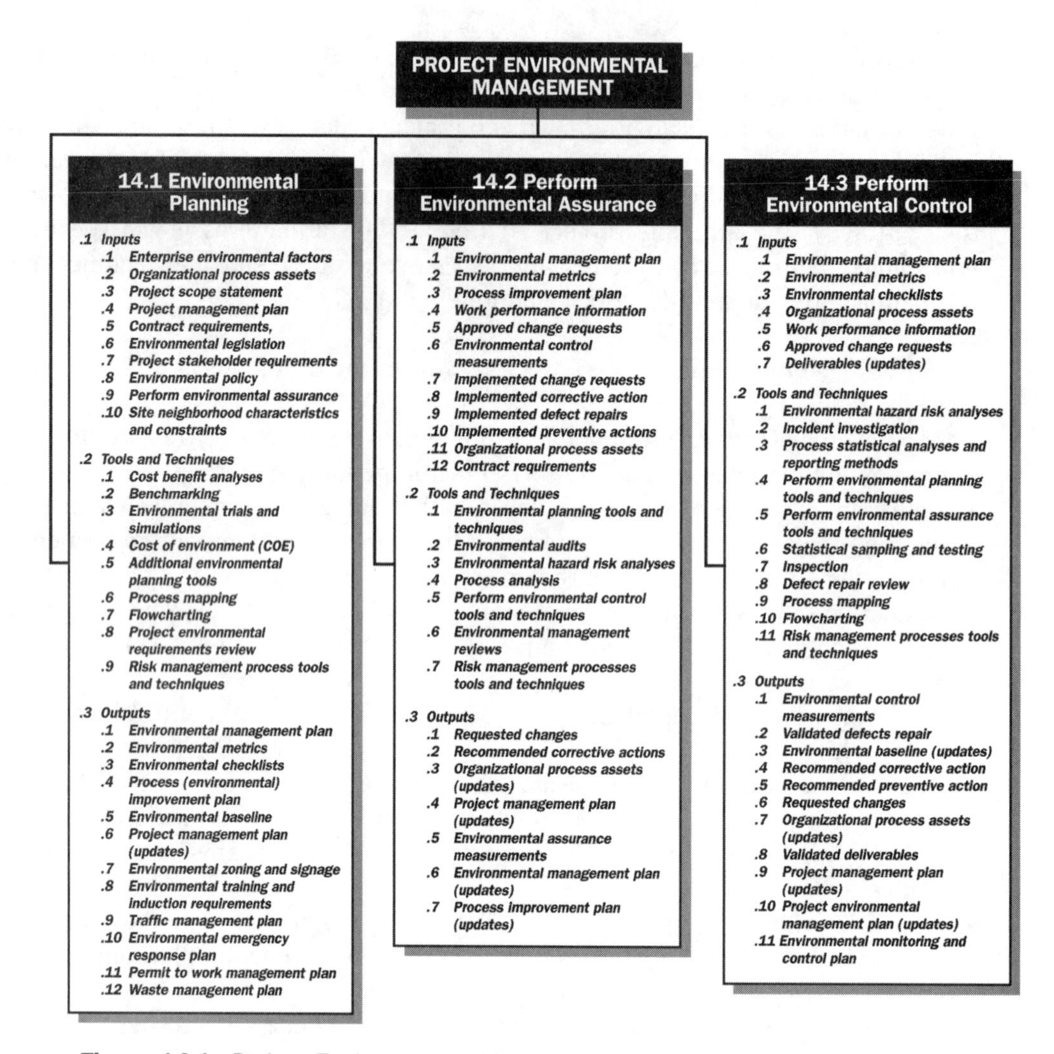

Figure 14-1. Project Environmental Management Overview

It should be noted that the lack of a particular environmental management program or system does not necessarily mean that the system employed by the performing organization is ineffective. Likewise, having an environmental management system or program does not necessarily signify that the performing organization will produce environmentally compliant products or work.

14.1 Environmental Planning

Environmental planning involves:

- Determining how to approach, plan, and execute the requirements for project environmental management.
- Determining the applicable requirements (environmental standards, regulations, specifications, etc.), which define the criteria to be used to determine both suitability of the project management system in fulfilling the requirements of the project and the acceptance of the product or the project.
- Assessing how best to apply the applicable acceptance criteria; documenting their characteristics and associated risks; and determining how to satisfy them.
- Determining how the suitability and effectiveness of project environmental management will be assessed or demonstrated.

Environmental standards incorporate project codes, regulations, and standards and they also include any condition of contract for which tangible deliverables have been defined and which will be used to determine environmental acceptance and compliance.

Environmental planning is one of the key processes when using the Planning Process Group (Section 3.2.12 of the *PMBOK® Guide*—Third Edition) for the development of the project management plan (Section 4.3 of the *PMBOK® Guide*—Third Edition and this extension), and these should be performed in parallel with and as part of other project planning processes. Critical to the development of an effective project environmental management system is an assessment as to how the performing organization will not only approach environmental management, but also plan and execute the project to meet the requirements for the environment. For example, the changes in the product required to meet identified or agreed-upon requirements (environmental standards, regulations, specifications, etc.) may dictate cost or schedule adjustments, or the desired product compliance may require a detailed risk analysis of the identified problem. One of the fundamental views of modern environmental management is that environment management must be planned, and built into processes, that is, environmental requirements must be identified and built into the manner in which processes and activities are carried out.

14.1.1 Safety Planning: Inputs

.1 Enterprise Environmental Factors
See Section 4.1.1.3 of the *PMBOK® Guide*—Third Edition.

.2 Organizational Process Assets
See Section 4.1.1.4 of the *PMBOK® Guide*—Third Edition.

.3 Project Scope Statement
See Section 5.2.3.1 of the *PMBOK® Guide*—Third Edition.

.4 Project Management Plan
See Section 4.3 of the *PMBOK® Guide*—Third Edition.

.5 Contract Requirements
See also Section 8.1.1.5 of this extension. Specifications, regulations, legislation, and standards (technical or legislative) are contractual requirements specific to construction projects.

Some construction projects may have additional requirements due to their nature, complexity or specific application area. For example, there are mandatory application area-specific environmental standards for construction within specific industries, for example nuclear projects, oil and gas projects, onshore and offshore projects, airport projects, or military projects, etc. In the construction industry, these requirements, issued by the project sponsor or owner, include the project scope statement and a description of the product(s) of the project, and make reference to all applicable standards and regulations.

.6 Environmental Legislation

Environmental legislation includes those requirements imposed by local, state, or federal government, and/or application area regulators. The may not always be stated in contract documents.

.7 Project Stakeholder Requirements

See Section 2.2 and Chapters 8 and 10 of the *PMBOK® Guide*—Third Edition.

.8 Environmental Policy

The environmental policy, although similar to the quality policy described in Section 8.1.1.1 of the *PMBOK® Guide*—Third Edition, differs in that it dictates much of the way in which construction activities are carried out from an environmental perspective. The environmental management policy also includes the degree to which the performing organization's management is committed to environmental issues and can have a major impact on the effectiveness of an environmental program.

.9 Perform Environmental Assurance Measurements

Perform environmental assurance measurements is the result of perform environmental assurance activities that are fed back into the Environmental Planning Process (Section 14.2.1 of this extension) for use in re-evaluating and analyzing the performance of the planning activities of the performing organization, the standards and processes employed, and where improvements are required. Perform environmental assurance measurements are also used as input into Perform environmental control (Section 14.3 of this extension) as an indicator of areas, which may need further investigation and to re-assess the risks or decisions taken in early project phases.

.10 Site Neighborhood Environmental Characteristics and Constraints

The characteristics of a construction site and its surrounding environment must be identified prior to project execution. The site neighborhood may fall under the authority of various constraints pertaining to environmental management, quality management, and safety management. These can include proximity of adjacent residents, configuration of project offices, layout and location of construction equipment workshop, material delivery time constraints, traffic congestion in vicinity of the project site during peak periods, water courses in the vicinity of the project, etc.

14.1.2 Environmental Planning: Tools and Techniques

.1 Cost-Benefit Analyses

See Section 8.1.2.1 of the *PMBOK® Guide*—Third Edition.

.2 Benchmarking

See Section 8.1.2.2 of the *PMBOK® Guide*—Third Edition.

.3 Environmental Trials and Simulations

Trials and simulations are discussed in Section 8.1.2.3 of the *PMBOK® Guide*—Third Edition. Examples include simulations of emergency responses procedures to environmental incidents to assure that the controls developed are adequate to address those incidents identified as requiring an emergency response, and are very much dependent on industry application area constraints and requirements (for example, mining, oil and gas, etc.).

.4 Cost of Environment (COE)

Cost of environment (COE) is a similar process to the cost of quality described in Section 8.1.2.4 of the *PMBOK® Guide*—Third Edition. Determination of the COE usually includes some form of cost-benefit analysis (see Section 8.1.2.1), an important consideration since environmental management is of paramount importance regardless of cost.

.5 Additional Environmental Planning Tools

Additional Environmental Planning Tools are similar to the additional quality planning tools described in Section 8.1.2.5 of the *PMBOK® Guide*—Third Edition.

.6 Process Mapping

Process mapping is commonly undertaken with flowcharting (see Section 13.1.2.7) to:

- Map how a particular process is carried out.
- Determine how various processes interact.
- Identify any gaps in a particular work item or activity (termed gap analysis), including the absence of critical review points or a required deliverable, such as omitting verification that work has been undertaken and is acceptable. Such verification could relate not only to quality, but also safety or environmental concerns.

.7 Flowcharting

See Section 8.3.2.3 of the *PMBOK® Guide*—Third Edition. A flowchart is a diagram that shows how various elements of a system or process (or series of processes) interact or interrelate. Flowcharting is commonly used with process mapping (see Section 13.1.2.6 of this extension) and with certain process statistical analyses and reporting methods. Flowcharting is also frequently used as a means of identifying non-value-added activities or functions, finding delay points in task activities or defining control points in work execution, for example, the issuance of a permit to enter prior to entering a confined space.

.8 Project Environmental Requirements Review

Project environmental requirements review includes an assessment and determination of the following:

- Characteristics and criteria of each activity and product of the project, and how to satisfy these. These are sometimes incorporated into activity risk assessments;
- Applicable verification criteria, including those required to demonstrate that acceptance and performance characteristics have been fulfilled;
- Alternatives review and selection. In construction projects, it is not uncommon for some activities to be potentially performed with multiple processes or arrangements for achieving the same result or output. This applies equally to safety management and quality management. Each process will require specific environmental requirements. Examples include the following:
    - Rock formations can be removed by blasting or by using pneumatic breakers.

   - Effluents from chemical pipe cleaning can be treated at an on-site waste treatment works, or taken to an external treatment facility.
   - Inspections can be carried out by independent inspection organizations (either in-house or external) or performed by those carrying out the work (provided the competence to do so has been determined, assessed and agreed upon).

  For each of these alternatives, the activity risk assessments would identify fundamentally different risks to workers.

Another example, (and one that has seen significant increase in today's global economy) as with quality management, is when a requirement (standard or specification) that is developed in one geographical region is used in another. In such instances, it is common for the characteristics and criteria of one region to differ in some degree from those in another location, as such requirements generally reflect the constraints in the region of origin. This is a case where compromise may be necessary, and require the re-qualification of the requirements to ensure compliance. The compromise does not imply lowering standards for quality, safety, or environment, but illustrates that the same end result can be achieved by different methods. This needs to be carefully scrutinized so as not to compromise quality, safety or environment, valid justification should be provided to project sponsors or owners, for obvious reasons.

Generally, all processes are analyzed to determine alternatives to increase effectiveness and efficiency, for example, cost-benefit analyses (see Sections 8.1.2.1 of the *PMBOK® Guide*—Third Edition) and analyses in which time, cost, quality, safety, and environmental aspects need to be balanced or even exceeded. Environmental requirements are often mandatory constraints, as non-compliance can cause execution permits to be cancelled, revoked, or not issued. Furthermore, a failure in any aspect of environmental management, can also manifest itself in more significant failures in quality or safety management.

- **Project stakeholder requirements review.** The requirements of project stakeholders is described in Section 2.2 of the *PMBOK® Guide*—Third Edition, and Section 8.1.2.8 of this extension, with the management of project stakeholder information described in Section 10.4 of the *PMBOK® Guide*—Third Edition.
- **Value engineering (VE).** Value engineering is a creative approach used on many construction projects to optimize product life cycle costs, save time, increase profits, improve environmental compliance, expand market share, solve problems, and use resources more effectively, etc., and is frequently undertaken in conjunction with other environmental planning tools and techniques, such as cost-benefit analyses, and design by experiments, etc.

.9 Risk Management Processes Tools and Techniques

Risk management processes tools and techniques are described in Chapter 11 of the *PMBOK® Guide*—Third Edition. They can be used to identify, qualify and quantify, quality, safety and environmental risks, and support response planning. Risk management processes tools and techniques should be used in conjunction with all other tools and techniques described previously to ensure an integrated approach to quality, safety, and environmental management. The information to be used as input is available from the inputs of Risk Management Planning, described in Section 11.1.1 of the *PMBOK® Guide*—Third Edition.

14.1.3 Environmental Planning: Outputs

.1 Environmental Management Plan

The project environmental management plan essentially defines the strategy or methodology to be adopted by the performing organization to undertake environmental management and to fulfill the requirements of the project, and is therefore a subset of the overall project management strategy or methodology and project management information system as described in Sections 4.1.2.2 and 4.1.2.3 of the *PMBOK® Guide*—Third Edition. The project environmental management plan is a high-level strategic planning deliverable that defines the overall intentions and direction of the performing organization as expressed by top management. This plan is reviewed at various stages throughout the project to ensure the continued ability to execute current and future phases of the project, and may involve the project sponsor/owner, and other major project stakeholders, for example, industry regulators and local, national, and federal government. The project environmental management strategy may contain information such as the following:

- **Environmental management staffing or human resources plan.** A subset of the project human resources plan developed as part of Section 9.1.3.3 of the *PMBOK® Guide*—Third Edition, developed by determining the various human resources arrangements, analyzing the risks and benefits of each, and selecting the optimum arrangement. The selected option needs to take into consideration the nature of the work involved and necessary competencies required, contract and legislative requirements, responsibilities and accountabilities, organization structure, interfaces/structures between the performing organization, project sponsor/owner and other project stakeholders, including the apportioning of work so as not to overload one particular function.
- **Project environmental management policy.** Described in Section 14.1.1.8 of this extension.
- **Environmental management budget.** A subset of Project Cost Management described in Section 7 of the *PMBOK® Guide*—Third Edition. The environmental management budget is developed by determining the different costs associated with the different approaches envisioned for the work to be carried out (including human resource arrangements), analyzing the costs associated with each and determining the optimum costs/budgetary requirements for environmental management.
- **Environmental records and documentation requirements.** The cornerstone of any environmental management system is the records and associated documentation generated and employed not only as the basis to determine satisfactory or unsatisfactory performance, but also the effectiveness of the project management system as a whole. The requirement for records and documentation is a subset of the requirements described in Chapter 10 on Project Communication Management of the *PMBOK® Guide*—Third Edition. Environmental management records and supporting documentation are also employed when assessing compliance with the requirements of statutory environmental requirements and legislation.
- **Agreed project stakeholder environmental requirements.** The agreed project stakeholder environmental requirements constitute an input into Project Stakeholder Management described in Section 10.4 of the *PMBOK® Guide*—Third Edition, and include project stakeholders' requirements for environmental planning, environmental assurance, and environmental control.

- **Environmental reporting requirements.** The reporting requirements for environmental management are a subset of the overall project performance reporting requirements described in Section 10.3 of the *PMBOK® Guide*—Third Edition. Environmental management reporting requirements would include, but are not limited to, assigned environmental management staff and resources, environmental baseline, environmental management planned vs. actual expenditure, agreed-upon performance and acceptance criteria, audit schedule, audits undertaken vs. those planned, details of audits (including periods for addressing non-satisfactory performance, details of corrective and preventive actions, statistical measurements to demonstrate project environmental efficiency and effectiveness of the project management system, etc.).
- **Project execution constraints.** The project execution constraints imposed must be taken into account, and may require additional mandatory quality, safety, and environmental requirements. Examples of such constraints are treatment of groundwater, working with noisy machines, and removing vegetation/debrushing, etc. At times, project execution characteristics may be dictated by the configuration of the construction site and that of the surrounding environment. Project execution constraints can often influence the approach or strategy adopted for quality, safety, or environmental management, and therefore form an input into Project Risk Management (Refer to Chapter 11 of the *PMBOK® Guide*—Third Edition).
- **Agreed environmental performance and acceptance criteria.** Criteria for the environmental performance and acceptance of all attributes of a project must be finalized and agreed upon with the project sponsor or owner and, where applicable, any project stakeholder. These criteria form the basis for the environmental management baseline (see 14.1.3.5).
- **Project administrative and contract closure procedures.** See Sections 4.7.3.1 and 4.7.3.2 of the *PMBOK® Guide*—Third Edition. Administration and contract closure include the collation and assembly of all pertinent environmental management records and other supporting documentation necessary to demonstrate that the environmental requirements of the project have been fulfilled.
- **Operational definitions.** The operational definitions describe, in specific terms, what something is and how it is to be measured. Quality, safety, or environmental control processes are frequently employed to determine how processes and project work is to be measured. For example, it is insufficient to state that management reviews of the project management system will be undertaken once per year; site environmental inspections will be carried out as per the frequencies specified in the contract; the hazard caused by construction equipment will remain within the permissible limits; or all site staff will wear personal protective equipment. The project management team must, in relation to these statements, be able to demonstrate and support that management reviews were performed, specific inspections were carried out, any oil spills or gas emissions were measured; whether only heavy vehicles or all vehicles were inspected; or that all site personnel were issued with personal protective equipment.
- **Environmental communication.** This type of communication can cover a broad range of activities such as:
  - Barriers, signs, and bulletin boards (see Section 14.1.3.7 of this extension);
  - Initial environmental indoctrination meetings, tool box meetings, individual bulletins on a specific subject, etc. (see Section 14.1.3.8 of this extension);
  - Environmental reports, including those required by legislation.

.2 Environmental Metrics

Environmental metrics are similar to quality matrices as described in Section 8.1.3.2 of the *PMBOK® Guide*—Third Edition

.3 Environmental Checklists

Environmental checklists are similar to quality checklists described in Section 8.1.3.3 of the *PMBOK® Guide*—Third Edition

.4 Process (Environmental) Improvement Plan

See Section 8.1.3.4 of the *PMBOK® Guide*—Third Edition.

.5 Environmental Baseline

See Section 8.1.3.5 of the *PMBOK® Guide*—Third Edition.

.6 Project Management Plan (Updates)

See Section 8.1.3.6 of the *PMBOK® Guide*—Third Edition.

.7 Environmental Zoning and Signage

Generally a project site can be divided into specific zones, each having its own particular environmental requirements (for example, workshops, storage areas, construction areas, limited or controlled access areas, etc.). The zoning of particular areas of a project site assists in determining the specific environmental hazards or risks associated with each. Consequently, the appropriate signage requirements forewarn construction personnel of specific hazards or controls to safeguard them. Such signage can include general signage to warn of access restrictions or more specific signage to warn of particular hazards. This process utilizes the principle of visual operations management (VOM), to reinforce and repeat any specific instruction that would have been given during environmental training and induction (see 14.1.3.8 of this extension). It is also common to use environmental signage banks at specific access points, where all necessary environmental signage relating to a specific area of the project would be stored and available for use, as needed. These banks would be subject to audit or inspection at regular intervals to ensure all appropriate signs are present and identify any that have been damaged or lost, and need replacement.

.8 Environmental Training and Induction Requirements

The requirements for environmental training are generally divided into the following categories:

- **Environmental training.** Environmental training is common for all construction projects, and includes mandatory statutory training (e.g., for operation of construction equipment) and performance of specific environmental tasks (e.g., sorting of different categories of waste). It has also become increasingly common that all construction personnel undergo mandatory general statutory environmental training before they are allowed onto a construction project site—a requirement resulting from the increased occurrence of environmental incidents which would, under normal circumstances, be seen as easily avoidable.
- **Environmental inductions.** Environmental inductions are also common on all construction projects, and relate to specific construction activities or tasks, to address the particular risks associated with these tasks and the necessary mitigation measures to be employed.

- **Tool box meetings.** Tool box meetings are generally undertaken to provide specific instruction regarding the use of certain construction equipment and tools, and would include such attributers as operation, maintenance, etc.

Critical to any environmental program is the need to plan when specific elements of the program must be implemented to address specific parts of the project. Environmental training planning is a subset of the human resources planning outputs (see 9.1.3 of the *PMBOK® Guide*—Third Edition).

.9 Traffic Management Plan

The traffic management plan defines the controls to be exercised over traffic in the vicinity of the project site, including, but not limited to, project-site entry and egress arrangements (including security checks), time limitations for deliveries, use of temporary roads for public traffic, weight restrictions, traffic signals and channeling of vehicles to avoid construction works, access and egress for emergency vehicles, etc.

.10 Environmental Emergency Response Plan

The necessity for a project-specific emergency response management plan is generally dictated by the constraints of the project and its environs, and such a plan would usually be developed in conjunction with project sponsors and owners. For example, underground works (such as tunneling) have specific mandatory requirements relating to emergency response needs. Requirements include, but are not limited to, responsibility and authority for key members of the emergency response team, communication requirements (especially of initial occurrence of emergency incidents, and with emergency services, etc.), provision of appropriate emergency response equipment, and access and egress requirements for emergency response vehicles (fire, police, ambulance, etc.). Emergency responses should not be limited to environmental incidents, and can also include safety emergency incidents. For example, inadvertent discharge of hazardous material can have both environmental and safety consequences. In addition, it is not uncommon for application area regulators to specify mandatory controls to be implemented as part of emergency environmental response activities.

It would be considered prudent to develop an integrated emergency response plan to address both safety and environmental incidents, especially as an emergency incident will directly or indirectly have safety and environmental implications

.11 Permit to Work Management Plan

The permit to work management plan is a subsidiary of the project management plan. Many construction project application areas have mandatory requirements relating to permits, for example, discharge permits (required to discharge into water courses), permit to excavate, and permit to enter biohazard areas, etc. (see 14.1.3.12). These permit procedures will generally be defined in conditions of contract and generally reflect the environmental checks that need to be addressed *prior* to commencing specific activities. This is especially true where part of the scope of work for a construction project relates to mitigation measures required to deal with contamination originating from adjacent industries. While the permit to work management plan is also a subset of the project communication management plan, the potential consequences of environmental failures warrant its placement in this chapter.

.12 **Waste Management Plan**

The project waste management plan is a subsidiary of the project management plan. Many, if not all, construction projects will produce some degree of waste, and some projects will generate significantly more waste than others (e.g., office development vs. infrastructure development). Project waste management generally employs the "3R process" to address environmental management (i.e. reduce, re-use and recycle), with the project waste management plan detailing the controls to be applied to each particular category of waste, ranging from general office waste (paper, etc.), food waste (from site canteens), sanitary waste (from site welfare facilities), through to the different categories of waste generated by construction processes, (excavations, metal, packaging, lumber, etc.).

Requirements for waste management include not only the identification of authorized waste disposal facilities, but also obtaining the necessary permits and authorizations (see 14.1.3.11) to discharge waste, the most common being:

- Discharge of wastewater during excavation into adjacent watercourse, where some degree of primary treatment is required, for example, water treatment plants, settlement ponds, etc., for the removal of suspended solids or other environmentally harmful constituent.
- Excavated material, where waste enforcement authorities are now empowered to assess the performing organization's controls for disposal controls and waste manifest system, that is, the determination of how much waste is placed in authorized areas.

With many construction projects now being sited in areas which would previously have been considered unsuitable, environmental management includes information on how to deal with possible contaminated areas. Such contamination includes material considered unsuitable for re-use or recycling, as well as contamination, which would constitute an environmental and/or safety hazard. No further discussion is made regarding this matter here, as it is beyond the scope of this standard in view of the involvement of local, state, and federal governments, statutory bodies, and industry regulators. It is mentioned here to make performing organizations aware that such instances are common. Additional information can be found in the *Government Extension to the PMBOK® Guide Third Edition.*

14.2 Perform Environmental Assurance

Perform environmental assurance involves the following:

- Applying the planned, systematic environmental activities to ensure that the project employs all processes needed to meet requirements.
- Determining whether these processes (and their integration) are effective in ensuring the project management system will fulfill the requirements of the project and the product of the project.
- Evaluating the results of environmental management on a regular basis to provide confidence that the project will satisfy the relevant environmental standards.

14.2.1 Perform Environmental Assurance: Inputs

.1 **Environmental Management Plan**

See Section 14.1.3.1 of this extension. The environmental management plan describes how environmental assurance will be applied and performed on the project, and defines the primary attributes of the project environmental management plan which form inputs to the perform environmental assurance process.

.2 Environmental Metrics
Environmental metrics are similar to quality matrices described in Section 8.1.3.2 of the *PMBOK® Guide*—Third Edition.

.3 Process Improvement Plan
See Section 8.1.3.4 of the *PMBOK® Guide*—Third Edition.

.4 Work Performance Information
See Sections 4.4.3.7 and 10.3.3.1 of the *PMBOK® Guide*—Third Edition.

.5 Approved Change Requests
See in Section 4.4.1.4 of the *PMBOK® Guide*—Third Edition.

.6 Environmental Control Measurements
See Section 14.3.3.1 of the *PMBOK® Guide*—Third Edition.

.7 Implemented Change Requests
See Section 4.4.3.3 of the *PMBOK® Guide*—Third Edition.

.8 Implemented Corrective Action
Described in Section 4.4.3.4 of the *PMBOK® Guide*—Third Edition.

.9 Implemented Defect Repairs
See Section 4.4.3.6 of the *PMBOK® Guide*—Third Edition.

.10 Implemented Preventive Actions
See Section 4.4.3.5 of the *PMBOK® Guide*—Third Edition.

.11 Organizational Process Assets
See Section 4.1.1.4 of the *PMBOK® Guide*—Third Edition.

.12 Contract Requirements
See Section 14.1.1.5 of this extension.

14.2.2 Perform Environmental Assurance: Tools and Techniques

.1 Environmental Planning Tools and Techniques
See Section 14.1.2 of this extension. These techniques are similar to the quality planning tools and techniques described in Sections 8.1.2 of the *PMBOK® Guide*—Third Edition.

.2 Environmental Audits
Audits are described in Section 8.2.2.2 of the *PMBOK® Guide*—Third Edition. Audits involve undertaking structured and independent reviews to demonstrate whether or not the project activities of the performing organization(s) comply with the project requirements.

Environmental audits of the product(s) of the project are termed environmental technical or environmental compliance audits and include an evaluation of the results or outputs of activities vs. the performance and acceptance criteria defined in technical environmental standards and specifications to determine fitness for purpose intended

Audits can also be undertaken for the project management system as a whole, or for individual component parts, for example, the procurement management system (see Section 12.5.2.3 of the *PMBOK® Guide*—Third Edition), design management system, or commissioning management system, etc. Audits are also carried out to assess compliance with statutory and legislative environmental requirements.

Integrated audits are commonly adopted, incorporating the applicable requirements such as those for quality, safety, and environmental management to afford a more accurate and comprehensive measure of the effectiveness of a specific area of work in fulfilling project requirements.

.3 Environmental Hazard Risk Analyses

Environmental hazard risk analyses are systematic reviews of a process, activity, or work element to identify the potential environmental hazards for project personnel as well as others who are present on the site associated with an activity or process. These analyses are normally carried out by specific and knowledgeable members of the project management team of the performing organization with the assistance of key construction supervisors and are part of the risk identification process outlined in Section 11.2 of the *PMBOK® Guide*—Third Edition.

.4 Process Analysis

Process analysis is described in Sections 8.2.2.3 of the *PMBOK® Guide*—Third Edition. Process analyses may also include statistical analyses (see 8.3.2.1 of the *PMBOK® Guide*—Third Edition).

.5 Perform Environmental Control Tools and Techniques

Perform environmental control tools and techniques is described as part of perform quality control tools and techniques in Sections 8.3.2 of the *PMBOK® Guide*—Third Edition, and Section 8.3.2 of this extension.

.6 Environmental Management Reviews

Environmental management reviews are discussed as part of quality management reviews detailed in Section 8.2.2.5 of this extension.

.7 Risk Management Tools and Techniques

Risk management tools and techniques are described in Section 11 of the *PMBOK® Guide*—Third Edition.

14.2.3 Perform Environmental Assurance: Outputs

.1 Requested Changes

See Sections 4.4.3.2 and 8.2.3.1 of the *PMBOK® Guide*—Third Edition.

.2 **Recommended Corrective Actions**
See Section 8.2.3.2 of the *PMBOK® Guide*—Third Edition.

.3 **Organizational Process Assets (Updates)**
See Section 8.2.3.3 of the *PMBOK® Guide*—Third Edition.

.4 **Project Management Plan (Updates)**
See Section 8.2.3.4 of the *PMBOK® Guide*—Third Edition.

.5 **Environmental Assurance Measurements**
Environmental assurance measurements are discussed as part of perform quality assurance measurements in Section 14.1.1.9 of this extension. Perform environmental assurance measurements is the result of perform environmental assurance activities that are fed back into the Environmental Planning Process (Section 14.1.1) for use in re-evaluating and analyzing the performance of the performing organization and the standards and processes employed. Perform environmental assurance measurements is also used as input into perform environmental control (see Section 14.3.1 of this extension) as an indicator of areas that may need further investigation.

.6 **Environmental Management Plan (Updates)**
See Section 13.1.3.1 of this extension

.7 **Process Improvement plan (Updates)**
See Section 8.1.3.4 of the *PMBOK® Guide*—Third Edition.

14.3 Perform Environmental Control

Perform environmental control involves the following:

- Determining and applying measures for monitoring the achievement of specific project results throughout the project to determine whether they comply with the requirements.
- Identifying unsatisfactory performance and identifying ways to eliminate causes of unsatisfactory environmental performance. This includes failures on the part of environmental planning and environmental assurance.

14.3.1 Perform Environmental Control: Inputs

.1 **Environmental Management Plan**
See Section 14.1.3.1 of this extension. The environmental management plan describes how environmental management will be applied and performed on the project. Section 14.1.3.1 defines the primary attributes of the project environmental management plan as follows:

- **Environmental management human resources plan.** See also Chapter 10 of the *PMBOK® Guide*—Third Edition). The organizational interface between the performing organization and project sponsor/owner, and other applicable project stakeholders is important for effective project environmental control.
- **Project environmental management policy.** See Section 14.1.1.8.

- **Environmental management budget.** See Chapter 7 of the *PMBOK® Guide*—Third Edition.
- **Environmental records and documentation requirements.** See also Chapter 10 on Project Communication Management. Environmental management records and supporting documentation (see 14.1.3.1) are also employed when assessing compliance with the requirements of ISO 14001.
- **Agreed project stakeholder environmental requirements.** See Section 10.4 of the *PMBOK® Guide*—Third Edition).
- **Environmental reporting requirements.** See Section 10.3 of the *PMBOK® Guide*—Third Edition and Section 14.1.3.1 of this extension.
- **Project execution environmental constraints.** See Sections 14.1.3.1 and 14.1.1.10).
- **Agreed environmental performance and acceptance criteria**. See Section 14.1.3.1).
- **Project administrative and contract closure procedures.** See Sections 4.7.3.1 and 4.7.3.2 of the *PMBOK® Guide*—Third Edition.

.2 **Environmental Metrics**
Environmental metrics are similar to the quality matrices described in Section 8.1.3.2 of the *PMBOK® Guide*—Third Edition.

.3 **Environmental Checklists**
Environmental checklists are similar to quality checklists as described in Section 8.1.3.3 of the *PMBOK® Guide*—Third Edition.

.4 **Organizational Process Assets**
See Section 4.1.1.4 of the *PMBOK® Guide*—Third Edition.

.5 **Work Performance Information**
See Sections 4.4.3.7 and 10.3.3.1 of the *PMBOK® Guide*—Third Edition.

.6 **Approved Change Requests**
See Section 4.4.1.4 of the *PMBOK® Guide*—Third Edition.

.7 **Deliverables (Updates)**
See Section 4.4.3.1 of the *PMBOK® Guide*—Third Edition.

14.3.2 Perform Environmental Control: Tools and Techniques

.1 **Environmental Hazard Risk Analyses**
See Section 14.2.2.3 of this extension.

.2 **Incident Investigation**
It is important that each incident be investigated as to the cause (direct or indirect), and a complete report should be made stating what happened and why. These reports are usually required by the insurance companies covering the resulting losses, but the reports are also vital as a measure of the performing organization's environmental performance and may help contribute to future improvements. In some cases, these reports are required by law enforcement agencies.

.3 Process Statistical Analyses and Reporting Methods

Process statistical analyses are described in Section 8.3.2 of the of the *PMBOK® Guide*—Third Edition. Common statistical analyses and reporting methods for environmental management include:

- **Cause and effect analyses**. See Section 8.3.2.1 of the *PMBOK® Guide*—Third Edition.
- **Control charts**. See Section 8.3.2.2 of the *PMBOK® Guide*—Third Edition and Section 8.3.2.2 of this extension.
- **Histograms**. See Section 8.3.2.4 of the *PMBOK® Guide*—Third Edition.
- **Pareto analyses.** See Section 8.3.2.5 of the *PMBOK® Guide*—Third Edition and Section 8.3.2.5 of this extension.
- **Run analyses**. See Section 8.3.2.6 of the *PMBOK® Guide*—Third Edition.
- **Scatter analyses and diagrams**. See Section 8.3.2.7 of the *PMBOK® Guide*—Third Edition.

In addition, for construction, the applicable statistical methods of analysis and their respective reporting methods are generally defined in applicable standards and specifications for specific project products, as described in Section 8.2.1.6 of the *PMBOK® Guide*—Third Edition.

.4 Perform Environmental Planning Tools and Techniques

See Section 14.1.2 of this extension.

.5 Perform Environmental Assurance Tools and Techniques

See Section 14.2.2 of this extension.

.6 Statistical Sampling and Testing

Statistical sampling is described in Section 8.3.2.8 of the *PMBOK® Guide*—Third Edition.

.7 Inspection

See Section 8.3.2.9 of the *PMBOK® Guide*—Third Edition and Section 8.2.3.10.

.8 Defect Repair Review

See Section 8.3.2.10 of the *PMBOK® Guide*—Third Edition. Although defects repair is generally considered to be an attribute of quality management (see Chapter 8 on Project Quality Management of the *PMBOK® Guide*—Third Edition), the review of defects with regard to any contributing environmental issues is not uncommon. Defects repair review is commonly addressed by means of the non-conformance control process.

.9 Process Mapping

See Section 13.1.2.6 of this extension.

.10 Flowcharting

See Section 8.3.2.3 of the *PMBOK® Guide*—Third Edition, and Section 13.1.2.7 of this extension.

.11 Risk Management Processes Tools and Techniques

Risk management processes tools and techniques are described in Chapter 11 of the *PMBOK ® Guide*—Third Edition.

14.3.3 Perform Environmental Control: Outputs

.1 Environmental Control Measurements

Similar to quality control measures described in Section 8.3 of this extension, these measurements represent the results of environmental control activities that are fed back as inputs into the Perform Environmental Planning (see Section 14.1.1 of this extension) and Perform Environmental Assurance (see Section 14.2.1 of this extension) processes of the performing organization to re-evaluate and analyze the performance of the project environmental management system, including the application of contract requirements (see Section 14.1.1.5 of this extension) and environmental legislation (see Section 14.1.1.6 of this extension).

.2 Validated Defects Repair

See Sections 4.4.1.6 and 8.3.3.2 of the *PMBOK® Guide*—Third Edition.

.3 Environmental Baseline (Updates)

See Section 14.1.3.5 of the *PMBOK® Guide*—Third Edition.

.4 Recommended Corrective Action

See Section 4.5.3.1 of the *PMBOK® Guide*—Third Edition and Section 14.3.3.4 of this extension.

.5 Recommended Preventive Action

See Section 4.5.3.2 of the *PMBOK® Guide*—Third Edition and Section 14.3.3.5 of this extension.

.6 Requested Changes

See Section 4.4.3.2 of the *PMBOK® Guide*—Third Edition and Section 14.3.3.6 of this extension.

.7 Organizational Process Assets (Updates)

See Sections 4.1.1.4 of the *PMBOK® Guide*—Third Edition.

.8 Validated Deliverables

See Section 14.3.3.7 of this extension.

.9 Project Management Plan (Updates)

See Section 4.6 of the *PMBOK® Guide*—Third Edition.

.10 Project Environmental Management Plan (Updates)

See Section 14.1.3.1 of this extension.

.11 Environmental Monitoring and Control Plan

The environmental monitoring and control plan describes how the project management team will implement the necessary environmental controlling activities of the performing organization. The environmental monitoring and control plan is a component or subsidiary plan of the project management plan (see Section 4.3 of the *PMBOK® Guide*—Third Edition).

The environmental monitoring and control plan will either contain or make reference to specific procedures to be employed to ensure the environmental compliance of the work being carried out. While the project environmental management plan details how the performing organization will undertake environmental management on the project, the project environmental monitoring and control plan defines the actual monitoring and control activities to be employed and undertaken, especially the following:

- Item of work to be monitored.
- Reference to the applicable reference document and acceptance criteria.
- Applicable verification activities (inspection, tests, reviews, submissions, etc.) which will be performed, and when such activities are performed in relation to the overall process.
- Responsibility for undertaking the work, and each verification activity.
- Applicable characteristics and measurements to be taken or recorded.
- Applicable supporting documentation to be generated to demonstrate satisfactory or unsatisfactory performance.

To be effective, environmental monitoring and control (and verification) must be integrated into the manner in which the physical work is performed. This integration process establishes control points or gates throughout the process to ensure that the next phase of work cannot proceed until the preceding work has been completed and verified as complete and compliant.

An emerging trend worldwide is the use of integrated execution and verification plans (EVPs) that incorporate (integrate) the necessary verification activities into actual work processes. EVPs also serve to plan work processes and the sequence by which such work needs to be performed. A similar approach can be adopted for monitoring and controlling activities for Quality Management (see Chapter 8 of this extension) and Safety Management (see Chapter 13 of this extension), with EVPs containing not only the precise sequence of work to be undertaken but also all necessary monitoring and controlling verification activities.

Chapter 15

Project Financial Management

Introduction to Financial Management

Financial management includes the processes to acquire and manage the financial resources for the project and, compared to cost management, is more concerned with revenue sources and monitoring net cash flows for the construction project than with managing day-to-day costs.

In traditional construction projects, the owner typically pays for the cost of the project by means of periodic (usually monthly) progress payments. The contractor thus only has to finance initial costs of set up and the first few months of work. Many contractors are able to these expenses or can obtain a short-term loan to cover this initial period.

More recently, however, the construction industry has faced increasing requirements to finance the entire project as a result of the use of several different types of project delivery methods. Some of these are design-build-own-operate (DBOO), design-build-operate-maintain (DBOM), lease-back, large projects with joint-venture partners, privatization of public projects, and projects that are non-recourse financed (that is, the project provides the sole collateral for the investors). This trend requires the contractor, who often leads any consortium involved, to be conversant with and somewhat knowledgeable about the subject and techniques of project financing.

Thus financial management is distinctly different from cost management, which relates more to managing the day-to-day costs of the project for labor and materials. In this section, the discussion is limited to financing the cost of construction of the project itself, although long-term financing may include both construction and operation, for example in the case of design-build-operate projects.

See Figure 15-1 for an overview of the processes and respective inputs, tools and techniques, and outputs. The major processes involved are as follows:

15.1 **Financial Planning.** Identifying key financial issues to be addressed and assigning project roles, responsibilities, and reporting relationships.

15.2 **Financial Control.** Monitoring key influences identified in Section 15.1 and taking corrective measures when negative trends are recognized.

15.3 **Administration and Records.** Designing and maintaining a financial information storage/retrieval database to enable financial control to proceed in a smooth way.

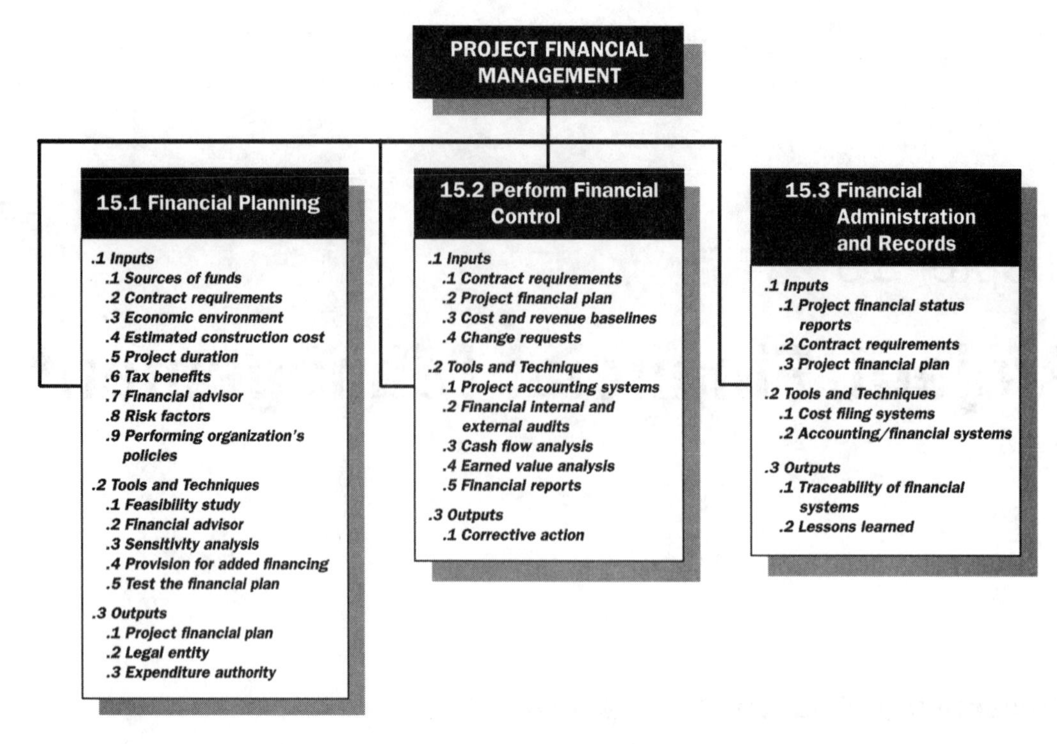

Figure 15-1. Project Financial Management Overview

These processes interact with each other and with the processes in other Knowledge Areas. Particular interaction is noted between the cost, risk, and time management Knowledge Areas. Each process may involve effort from one or more individuals or groups of individuals, depending upon the needs of the project.

While the processes are presented here as discrete elements with well-defined interfaces, in practice, they may overlap and interact in ways not detailed here.

15.1 Financial Planning

For construction projects, planning is the initial phase of financial management, as with any other Knowledge Area. Planning is the phase where all financial requirements are identified and provided for in the project. Financial planning is no different than standard project planning in that tasks are identified and requirements are quantified and placed on a timescale. Resources are also required to ensure that the financial tasks are completed on time.

15.1.1 Financial Planning: Inputs

.1 Sources of Funds

The sources of funds for a traditional project are often obtained from a company's central financing system, which may be a combination of borrowing from financial institutions, retained profits, financial reserves, and as noted in the introduction, down payments and progress payments by the client. The costs of financing are normally charged as interest to the project's cost account for construction projects.

For fully funded projects, there are many possible sources, such as commercial paper backed by a bank credit facility, bank loans, public debt offerings, private placements in the U.S. and European markets, syndicated commercial long-term loans, and government entity loans, etc. In some cases, funding may occur incrementally at different phases in the project. The funding points of the project need to be considered in the financial plan, as well as in the overall project plan, in order to maintain the momentum and the continuation of the project. Final determination of funding sources depends in large part on the project's credit worthiness and project sensitivity to changes in interest rates. In almost all of these types of projects, the participants acquire some equity in the project.

.2 Contract Requirements

For traditional projects, the client's financial status should be investigated to ensure that the client has the financial means to cover the project. If there is some doubt, it may be possible to obtain an irrevocable letter of credit. It is of prime importance to ensure that the client can service all agreed-upon payments.

For projects to be contractor financed, the contract may contain important clauses that restrict the contractor's ability to obtain favorable terms. Since this type of project is often awarded after a proposal process, there may be an opportunity to negotiate more favorable terms.

The contract and the project plan will help define requirements for the financing needs of construction projects. The contractual terms of payment from the client are utilized as input in ascertaining the financial needs of a project as this will help in the estimation of the cash flow, which will influence the project finances.

All financial items impact the bottom line of the project and must be scheduled within the overall costs of the project. Costs that should be taken into account include the following and should be assigned at least one level within the WBS as follows:

- Currency hedging (if payment currencies are different from purchasing currencies or visa versa)
- Costs of bonds and bank guarantees such as performance/down payment/warranty bonds
- Costs of interest on borrowings (should the project run with a negative cash flow).

It is also very important in volatile markets to consider price variations like interest rates and escalation of costs, especially on long-term construction projects, where prices can increase dramatically over the running time of a project.

.3 Economic Environment

The economic environment is an external factor that is not in a project manager's control, but the project manager must be aware of all the risks in this area and must periodically ensure that the financial plan is updated to allow for these risks. Factors may include political, regulatory, social, and economic factors such as currency fluctuation, which can increase or decrease the cost of the project.

.4 Estimated Construction Cost

Lending institutions will examine carefully the estimated cost of a project before committing to lending or participating in the project.

.5 **Project Duration**

All participants are affected by the length of a project, which will determine the length of their investment and when they will recover it, together with any expected profit.

.6 **Tax Benefits**

Many long-term projects may provide tax benefits that need to be taken into account when arriving at the financial plan.

.7 **Financial Advisor**

It is strongly advised for a contractor involved in a fully funded project to consult with an experienced financial advisor who has access to and knowledge of public and private capital markets.

.8 **Risk Factors**

A proper financial plan will allocate risks among participants, investors, customers, and interested third parties. Some of the risks most relevant to obtaining favorable financing are completion risk, cost overrun, regulatory and political risk, and technology risk (see Section 11.1.2). Further, it is not uncommon for a financial institution to ask for a complete risk analysis of the project's potential environmental impacts in order to assure that funds will be adequately applied for minimizing environmental impacts.

.9 **Performing Organization's Policies**

An organization's policies and strategies have influence over how a project is priced and should be considered during the financial plan (see Section 4.1.1.4 of this extension).

15.1.2 Financial Planning: Tools and Techniques

.1 **Feasibility Study**

For long-term projects financed by the contractor, a study needs to be conducted to determine if the project can be profitable within the given parameters. When proposing, a study needs to determine whether the ultimate payments proposed will cover all of the costs and still provide a reasonable profit.

Cash-flow measurement is a prime way of determining the viability of a project. Construction projects rely on cash inflow to balance out the costs incurred in order to keep financial costs to a minimum. Based on the terms and conditions of payments for the construction contract, a project manager can identify when and how money inflow will occur.

Money outflow refers to the scheduled payments for the subcontractors, vendors, and fees, insurance, taxes, direct labor, and support staff, including the cost of financing. The overall project plan provides the information to estimate what the periodic expenditures would be over the project's life cycle. This outflow schedule provides the needed information on the financial requirements for each period.

By analyzing revenue and expenditure, the net cash flow (inflow minus outflow) and basic financing requirements can be determined.

.2 Financial Advisor

A financial advisor participating in this part of the process assumes the principal responsibility for developing a comprehensive marketing strategy that will implement the financing plan in an optimum manner. Lending institutions and other fund sources are contacted and requested to provide part of the financing required.

.3 Sensitivity Analysis

A sensitivity analysis should be performed, which examines variances among several parameters, to determine the effect upon the project's cash flow and the preliminary financing plan.

.4 Provision for Added Financing

A study should be conducted to determine potential added financial needs, covering such items as unexpected delays, scope revisions, and other risks, so that the financing plan can provide for additional financing, if necessary.

.5 Test the Financial Plan

It is good practice to test the financial plan by contacting prospective lenders to ensure the acceptability of any unique features in the financing plan.

15.1.3 Financial Planning: Outputs

.1 Project Financial Plan

With the previously mentioned inputs, tools and techniques, a comprehensive financial plan can be developed which would clearly identify all financial requirements of a construction project and the means to finance them. All parties must understand when and by whom all of the necessary equity, debt, and insurance, in appropriate types and amounts, are to be supplied during the construction period.

.2 Legal Entity

The participants and the financial advisor must decide upon the legal form of the venture that would be most appropriate and advantageous, such as, partnership, corporation, trust, joint venture, or a combination thereof.

.3 Expenditure Authority

Authority for expenditure by the project manager is usually determined by the company commitment policy and, preferably, is built into the organizational structure. It is normal that certain levels within a project are authorized to spend a specified amount before an additional signature is required. It is wise for all financial aspects to require dual signatories, but the delegation of authority also depends upon the size and the nature of the project. In the case of a newly formed entity, these authorities need to be incorporated into the legal structure of the new entity.

15.2 Perform Financial Control

Financial control ensures that bonds are reduced when necessary, calls for funds from project partners are made as needed and all insurance and bank withdrawals/deposits are performed at the appropriate times. Financial control and cost controls are executed in the most effective way to ensure all items are within budget and the financial cash forecast.

15.2.1 Perform Financial Control: Inputs

.1 Contract Requirements

See Section 15.1.1.2 of this extension.

.2 Project Financial Plan

See Section 15.1.3.1 of this extension.

.3 Cost and Revenue Baselines

The budget and revenue forecasts developed for the financial plan serve as the net cash flow baselines for the project, against which any adjustments are measured.

.4 Change Requests

Any changes to the project that affect cost or revenue streams must be analyzed and incorporated into the financial plan to determine their effect on long- or short-term borrowing, insurance coverage, and other features of the financial plan.

15.2.2 Perform Financial Control: Tool and Techniques

.1 Project Accounting Systems

The project accounting system should be similar in structure to the WBS, showing the breakdown of the total project in more controllable modules. On small- to medium-size projects, the breakdowns can be kept on simple Microsoft Excel generated S curves. However, for large projects, accounting systems are usually more sophisticated. Financial control is exercised by closely monitoring actual spending and revenue against budget and cash flow forecasts, adjusting either the work methods or problem areas where this mechanism shows deviations.

.2 Financial Internal and External Audits

Internal and/or external audits ensure correct accounting methods and financial practices are being maintained. These audits are often very helpful to the project manager in uncovering problems that otherwise might not be seen. External audits are often a statutory requirement of the local government.

.3 Cash Flow Analysis

Updating all of the actual financial and cost data regularly provides an up-to-date financial information system from which the project manager analyzes trends in the system based on unique characteristics of the project. From these trends and past actual data, the project manager can revise the forecast for the remaining duration.

.4 **Earned Value Analysis**
Utilizing cost and schedule updated information (See Chapters 6, 7 and 10), an analysis of cash flow trends and forecasts can be performed to determine what adjustments may be required to the financial plan.

.5 **Financial Reports**
For projects, that need full financing, periodic financial reports are required by management and any lenders involved. When projects are comprised of some form of consortium or partnership, periodic (often monthly or quarterly) meetings are typical, during which project leaders present the status of the project and forecast its future, including the state of its financial health.

15.2.3 Perform Financial Control: Outputs

.1 **Corrective Action**
Based on the project health, financial status, and an analysis of financial status against established criteria, typically an action plan is prepared to correct the original forecasts and plan. Budgets may need to be revisited and adjusted according to the current state of the project, which may require approval from the participants. There may also be a need to increase revenue from financial sources to cover any projected shortfalls.

15.3 Financial Administration and Records

15.3.1 Financial Administration and Records: Inputs

.1 **Project Financial Status Reports**
See Section 15.2.2.5 of this extension

.2 **Contract Requirements**
Contract clauses need to be reviewed for invoicing and possible bond reductions that may require a certificate of tax paid and specific statements of completion. Some contracts require written indemnities for the client, certification of salaries paid to employees, and subcontractors for payment of progress invoices. All requirements must be complied with to avoid delays in payments and serious interruption of project cash flow.

.3 **Project Financial Plan**
See Section 15.1.3.1 of this extension.

15.3.2 Financial Administration and Records: Tools and Techniques

.1 **Cost Filing Systems**
See Section 15.2.2.1 of this extension.

.2 **Accounting/Financial Systems**
See Section 15.2.2.1 of this extension.

15.3.3 Financial Administration and Records: Outputs

.1 Traceability of Financial Systems

Traceability/retrieval within financial systems is very important to allow auditors and company management to evaluate the company's and the project's financial health. Traceability is easy if the storage of financial information is well defined and standardized. Computer-aided financial information storage is much more traceable and less time-consuming, compared to the conventional filing and retrieval system.

.2 Lessons Learned

Financial records and reports illustrate the problem areas previously encountered and corrective actions taken. For projects with long durations, lessons learned from past experience can be valuable in avoiding similar problems over the future life of the project.

Chapter 16

Project Claim Management

Claim management describes the processes required to prevent construction claims, to mitigate the effects of those that do occur, and to handle claims quickly and effectively. Although agreed-upon changes to the contract documents occur frequently, disputes among the stakeholders of a project are almost as common. To a certain extent, contract claims are addressed in Section 12.5.2.6 on Claims Administration of the *PMBOK® Guide*—Third Edition and this extension. This chapter was added to provide a general standard for addressing claims, because dealing with claims is such a significant part of the construction process.

Claims can be viewed from two perspectives: the party making the claim and the one defending against it. A claim is a demand for something due or believed to be due, usually as a result of an action, direction, or change order against the agreed terms and conditions of a contract and that can not be economically resolved between the parties. In construction, the demand is usually made for additional compensation for work claimed to be outside of the contract, or an extension of time for completion, or both. There seems to be endless opportunities for claims to take place. However, the distinction between a claim and a change is the element of disagreement between the parties as to what is due and whether or not anything is due. If agreement is reached, then the claim disappears and becomes a change. If not, the claim may proceed to negotiation, mediation, arbitration or, finally, to litigation before it is ultimately resolved. A claim is a contractual means to resolve a time and/or cost issue. Often claims are thought of in terms of a contractor making claims against an owner or other prime party and by subcontractors against a contractor; but claims can also originate with an owner who believes that some requirement of the contract is not being performed by the contractor. Unresolved issues can escalate into a claim and become a fierce contractual dispute among the stakeholders.

This chapter presents an outline for claim management, including appropriate dispute resolution methods. This chapter is intended to stimulate a careful approach to contract preparation, contract management, project documentation and expeditious handling of claims, when applicable.

See Figure 16-1 for an overview of the processes and respective inputs, tools and techniques, and outputs. Claim management which is applicable throughout the life cycle of a construction project, consists of the following four processes:

16.1 Claim Identification

16.2 Claim Quantification

16.3 Claim Prevention

16.4 Claim Resolution

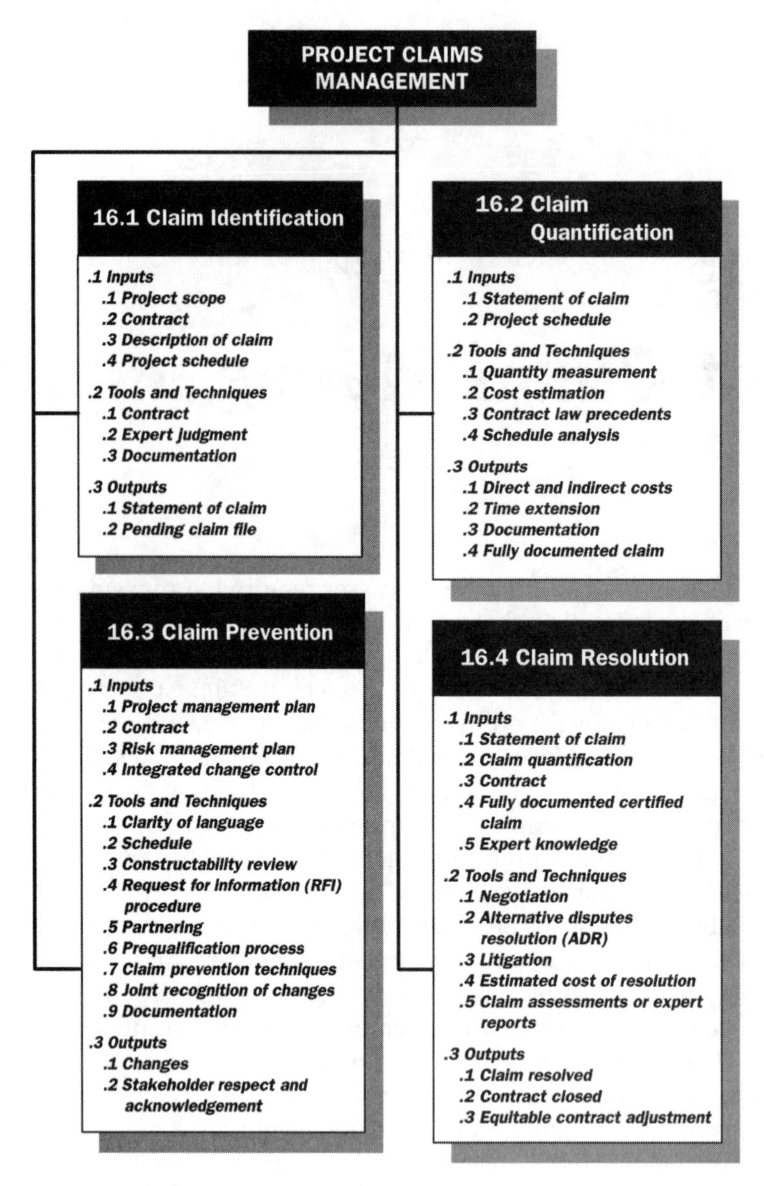

Figure 16-1. Claim Management Overview

16.1 Claim Identification

From the stakeholder's perspective, the goal for claim management is to prevent claims entirely and, if not, to resolve them at the earliest opportunity for the least cost and least disruption to the project. To do so, one must first possess the capability to recognize potential claim situations, either contractual or performance-based. Therefore, identification of a claim starts with sufficient knowledge of the project scope and contract requirements, which leads to an awareness that some activity may involve a change in scope or the contract. Proper identification initially requires an interpretation of the contract documents, followed by a documented description of the activity viewed as beyond these requirements.

16.1.1 Claim Identification: Inputs

Most claims are the result of an unresolved request for a change order or are a result of a contract document interpretation. Sometimes, a contract is not administered in a timely manner, causing time/cost impacts on the project. Project performance is another area where claims may arise due to either the quality of work performed or the progress of the work not being obtained. Regardless of the origin of the potential claim, claim identification must start with an emphasis on preventing claims and this effort begins with inputs to claim identification.

.1 Project Scope

See Chapter 5 of the *PMBOK® Guide*—Third Edition, which covers the scope of work as set forth in the contract, including all plans and specifications.

.2 Contract

The contract generally includes the various terms and conditions that apply to the work, provisions relating to changes, changed conditions, schedule preparation and submittal, and appropriate notice requirements.

.3 Description of Claim

A written description of the work believed to be outside of the contract, where it occurred, and when it took place. A statement of why it is not covered in the contract scope and reference to the section of the contract that supports the contention should accompany the claim. It is also important to document the activities that were affected and who performed the work. The claim should also address the project schedule and the additional contract time, if required, to perform this extra work. The contemporaneous documentation should be well recorded (see Chapter 10 on Project Communications Management and Section 10.3.2 on Tools and Techniques for recording project performance).

.4 Project Schedule

Should the claim impact the contract time, a record of when the alleged extra work commenced or will commence, and when it ended or is estimated to end should be documented. The project schedule becomes one of the key records for demonstrating the impact on the project. Time extension claims which result from delays to the work due to events such as unusual weather, strikes, or other *force majeure* items outside of the contractor's control, may be valid although they may not be compensable.

16.1.2 Claim Identification: Tools and Techniques

.1 Contract

The contract consists of the provisions relating to changes and notice provisions. In many cases, claims not made in a timely manner are invalid. See also Section 12.5 on Contract Administration of this extension and Section 12.4.3.2 on Contract of the *PMBOK® Guide*—Third Edition.

.2 Expert Judgment

Depending on the project delivery method employed, the owner may seek advice from one or more of the stakeholders as to whether a claim is valid. This may be the designer of record or the construction manager on a construction management scenario. It is often worthwhile to reach a consensus among several people as to whether or not the activity under question merits claim status. In some situations, such as with expensive or technically challenging claims, legal advice and/or construction consultants experienced with claims may be consulted for expert opinions.

.3 **Documentation**

One of the most important factors in the claim process is the need for proper and thorough supportive documentation. Such documentation may take many forms, including photographs and videos of the work in question, relevant contract sections and drawings, relevant statements of persons involved in or related to the claim work. Any correspondence, instructions, approved details, or in-shop drawings related to the claim, are also considered to be supportive documents. In addition, as-built schedule information for the time and work days for performance of the claim should be noted. Moreover, the contract documents may require separate cost accounting for claimed extra work. See also 10.1.3.1 on Communications Management Plan and 10.2.2.5 on Manage Project Documentation in this extension, for generally recognized acceptable practices for project documentation.

16.1.3 Claim Identification: Outputs

.1 **Statement of Claim**

With the information gathered in Section 16.1.2 of this extension, it is possible to prepare and submit a complete statement of the claim in accordance with the process and procedures set forth in the contract.

.2 **Pending Claim File**

Items identified as potential claims should be kept in their own separate files that contain all related documentation, including the statement of claim, correspondence and contract references. This assures that relevant information and documents are collected and retained in a single location. See Section 16.1.2.3 of this extension which provides important information pertaining to documentation.

16.2 Claim Quantification

Once an item or issue has been reviewed and identified as a potential claim, a decision must be made to determine if a claim is worthy of pursuit. An adequate assessment should be made of the possible stakeholder consequences, positive and negative, when a claim is submitted. The next step is to quantify the potential claim in terms of additional compensation, a time extension to the contract completion, or both. Claims often tend to create barriers for prompt resolution due to the differences in the cost and time impact perspectives between the stakeholders. Nevertheless, there are proper and logical ways of determining the cost of the extra activity or damages, both in terms of time and money. The process basically uses a cause and effect approach to determine the full effect of the claimed extra work or delay activity and the full effect on the construction work caused by the claimed item. Sometimes the claimed item has an indirect or direct effect on other aspects of the construction project, making such works more costly, changing sequences, or delaying other activities. To the extent that these indirect effects can be justified and quantified, they are properly part of the total cost of the claim.

16.2.1 Claim Quantification: Inputs

.1 Statement of Claim

See Section 16.1.3 of this extension. An important aspect of the statement of claim that must be reviewed for claim quantification, is the specific component of the contract which prescribes cost items that either can or cannot be claimed. Jurisdictional or international laws regarding reimbursement may also affect the claim elements related to cost reimbursement.

.2 Project Schedule

In the event that a claim occurs, this is the additional effect or consequence on the balance of the contract work caused by the claimed activity. These effects should be treated in the same manner as data collected for the claimed activity itself. Careful and explicit documentation needs to be maintained to fully support these types of activity impacts, including the use of the project schedule to demonstrate the claim impact. See also Section 10.2.2.5 of this extension for work performance and contemporaneous documentation components.

16.2.2 Claim Quantification: Tools and Techniques

.1 Quantity Measurement

Actual quantities of the claimed work are measured in terms of cubic meters of concrete or earth, weight of steel, or linear measures of piping and electrical work, etc. When disagreements arise, agreement of the specified quantity is one of the first aspects to review. In addition to material quantification, detailed activity cost tracking can document all resource hours expended on the project. Actual labor hours can be quantified for the extra work or for the disruption of other activities. It is important that this quantification be accurately performed and treated as a separate line item in the cost report process. With effective labor tracking, the quantification of extra work hours is relatively easy and is a far more effective and credible method for substantiating actual costs than an after-the-fact, subjective description or estimation of labor. Further, not all extra work can be quantified by unit weight or measure, so the difficulty or complexity of work performance, is best documented by a time and material process.

.2 Cost Estimation

Cost estimation covers the cost of the labor, material, and equipment involved in the claimed work. If cost records are available, they will provide the basis of the estimate. If not, the cost will have to be estimated using current applicable rates. Additions for overhead and profit are common and usually proper, because, at this stage, the claim is treated as though it will be considered a change. Sometimes the claimed work has an affect on other work on the project that results in additional cost. Usually this cost will have to be estimated, since the cause-and-effect relationship is not obvious. Often, though, the justification for this indirect effect is difficult to prove to the satisfaction of the opposing party. In some situations, given quantifiable work performance, a measured-mile-cost approach, wherein, actual costs and work efforts are recorded for very similar work that was not affected by the claim, can be used to determine, through comparative analysis, the cost impact of disrupted and impacted work activities. In addition to direct project costs, the stakeholder's business can also be impacted by a delayed occupancy, by unavailability of contract labor and equipment resources to perform work on other projects, or additional or extended general administrative costs to manage the project.

Most contracts include a liquidated damages clause, should the project not be completed within the contract requirements, but the contractor can also be hurt by projects that limit its ability to perform other projects, current or prospective, and these damages are sometimes called lost opportunity costs.

.3 Contract Law Precedents

Relevant case law may provide guidance as to what may or may not be included in the claim or how the claim may be evaluated. There are times when it is appropriate to seek legal counsel to provide advice on relevant jurisdictional court rulings.

.4 Schedule Analysis

There are a number of types of schedule analysis. Some of the available, sophisticated computer programs can help with this analysis, but can also make it overcomplicated and can create illusionary effects. The common way of assessing the schedule effect of changes and claims is to compare the as-planned schedule with the as-built schedule to support the time extension requested. The best practice is to support the actual activity events with effective and consistent schedule documentation (see Section 6.6.2 on schedule control of this extension. The effect on the critical path of the project schedule can be difficult to isolate because of all of the factors which can and do affect construction schedules. Such delays to the project may be due to owner actions or inactions, for example, late or partial access to work areas, information, or decisions. Further, when delays are caused by both parties, a separate analysis may be required for each case in order to properly allocate project delay responsibility.

16.2.3 Claim Quantification: Outputs

.1 Direct and Indirect Costs

Direct and indirect costs include a report of the costs or damages resulting from the claimed activity, including the supporting factors used in the calculation. Also, the cost, when justified, of the effects of the claimed activity on other aspects of the construction project (i.e., indirect costs), calculated in the same manner as the direct costs.

.2 Time Extension

Time extension is the result from the analysis in Section 16.2.2.4 of this extension.

.3 Documentation

Proper data to support quantity calculations, including time cards showing the extent of labor involved and machine usage, wage rates, equipment rates, and invoices for materials that are included in the claim.

.4 Fully Documented Claim

A complete document that presents the quantification of the cost impact, the time extension request, and all supporting/substantiating documentation. Usually, the submission procedures for claims are set forth in the contract. In certain government contracts, claims must be certified by a senior executive of the business that has authority to make such certification. This document can provide an outline of the claim, along with the specific costs that are being claimed. The claim document may become the working document which the opposing party will use to define its defense position or acquiesce to the claimed costs.

16.3 Claim Prevention

Construction activities are generally carried out in complex, highly sensitive and changing environments. Perfect conditions and control are nearly impossible to obtain. Clearly, the best way to prevent claims is to eliminate claims. Thus, the emphasis is on how to keep issues from arising which could develop into claims. The perfect, well-scoped, risk-allocated, and well-executed contract is less likely to result in any claims. However, since perfection is seldom attainable, owners and contractors can only strive towards achieving that goal. Having a well-developed plan is critical, one providing for flexible implementation consistent with the plan, as well as thorough communications with all stakeholders. Lack of follow-through, random changes to performance or schedule, unknown conditions, resource problems, and slow decision making are all factors that can negatively affect the project. Most stakeholders are aware that the project will be challenged by such factors, but how one reacts to these conditions can either lead to or help prevent the claim. Early recognition of potential problems, and open communications regarding possible alternatives or changes to the plan, create a collaborative environment, in which claims are less likely to occur.

There are several general principles of practice described in this chapter, which if followed, can mitigate or even eliminate the occurrence of claims.

16.3.1 Claim Prevention: Inputs

.1 Project Management Plan

The fundamental parts of the plan are the most important to claim prevention. A clear and carefully described scope of work, a reasonable schedule, an appropriate method of project execution tailored to the type of project, and an acceptable degree of risk involved, contribute towards the elimination of claims. An assessment of the construction works encompassing all requirements of the project plan should be satisfied in conjunction with the major stakeholders to the project plan. See also 9.3.2.3 on team-building activities of this extension for opportunities to build consensus and collaboration among the team members as early as possible at the start of construction. See also 9.1.1.1 regarding project interfaces, for considerations that may have an effect on timely decisions at the site and at administrative levels. If partnering is to be implemented (see 16.3.2.5 of this extension), it is important that the stakeholders take it seriously and have a proactive attitude.

.2 Contract

Fairly drawn contract terms that provide for possible changes and unknown site conditions, *force majeure*-type delays, periodic reporting, fair-notice provisions, and approval times also provide a basis for minimizing claims. A review of the dispute resolution clauses prior to attempting to address a dispute can help alleviate uncertainty and ambiguity on these matters and keep the dispute from escalating to a full claim.

.3 Risk Management Plan

Claims are minimized by the use of a risk management plan that allocates the risk between the parties on the basis of which one has the most control over the risk factor involved. The common owner's practice of trying to hold the contractor responsible for maximum amounts of risk, even in areas where the contractor has little or no control, is an invitation to claims.

.4 Integrated Change Control

See Chapter 4 on Project Integration Management and Section 4.6 of this extension for generally recognized good practices for managing changes. In addition, an effective strategy for handling disputes is to avoid hard-lined positions which only serve to escalate a claim situation. The change control process should designate levels of authority for the decision-making process.

16.3.2 Claim Prevention: Tools and Techniques

.1 Clarity of Language

The contract scope and specifications should be written in clear, unambiguous terms. Sometimes a pre-tender meeting between the stakeholders prior to contract finalization will help to clarify the details, requirements, and expectations for the project. Declarations from the contracting parties regarding bid quantities can also prevent claims arising from project scope.

.2 Schedule

Requirements for cost, schedule, scope, and specifications should be clearly stated, reasonable to accomplish, and mutually agreed upon by stakeholders. Effective and qualitative reviews by all parties will increase clarity around the schedule, planned work, and progress. Scheduled update submission requirements should be fair and capable of providing good schedule as-built documentation without unneeded complexity.

.3 Constructability Review

The use of a constructability review can avoid later field errors and unnecessary changes in construction methods, all of which can lead to claims.

.4 Request for Information (RFI) Procedure

Contracts requiring designer or owner approval for shop drawings, materials of construction, RFIs, and similar items should contain a clause stating a reasonable time for the answer to be given. If this deadline is not met, the contractor may have grounds for a claim; but those grounds will be clear to all. Consequently, the time for the response should be realistic—not inordinately long nor so short that it will be difficult for the owner or designer to comply. See also Section 10.3.2 on Performance Reporting: Tools and Techniques of this extension, for other process and system tools that contribute to effective communication and reporting on project progress, events, and administrative processes, such as change order logs, change order proposals, submittal tracking and progress meetings.

.5 Partnering

Projects that use the technique of project-specific partnering are significantly less susceptible to the problem of claims because of the atmosphere of mutual dedication and commitment that such partnering promotes. Processes and systems can be established which lead to much better communication, timely turnaround of submittals, prompt decision-making, and often type of issue escalation process, when a potential issue is uncovered. Partnering efforts are not considered added requirements to the contract, but rather commitments of the stakeholders' parties to make it work and to make the project a success. One might consider this technique an alignment of the project management processes between the primary stakeholders.

.6 Prequalification Process

Projects that utilize prequalification of contractors have the benefit of dealing with seasoned and qualified contractors who are less likely to find themselves in desperate situations that might drive others to frantic claim activity.

.7 Claim Prevention Techniques

The construction industry has made great progress in creating alternative techniques for resolving contractual disputes. These are covered in Section 16.4.2 of this extension. However, proactive processes can also be implemented prior to resorting to the full claim process. The following list offers some of the more prevalent techniques that can be considered for use during the project and prior to the claim process. Several professional organizations exist globally that retain or have access to the professional resources that can provide these services:

- **Dispute review board (DRB).** Some projects, usually larger ones, establish a DRB at the outset of the project, with good results. The DRB acts as a kind of arbitration panel over any disputes that arise during the project so that potential claims are turned into changes or are dismissed for good reason before the project is ended.
- **Independent neutral.** On moderately sized projects, stakeholders can opt to share jointly in the cost of a third party, neutral to the project, who can act as a technical or construction advisor to the stakeholders. This neutral party can also analyze the situation and provide recommendations for fair solutions, including the time and cost impact assessment.
- **Intervention partnering.** Similar to the independent neutral, a partnering facilitator who is skilled in disputes and mediation can be brought in during the course of the project to facilitate an alternative resolution process. This process often involves independent experts that will assess the project, evaluate the claims, and assist the stakeholders in effectively completing the project and in mitigating consequential impacts and further delays.
- **Mediation.** While this is most often implemented at the close of a project with claims outstanding, nothing precludes the use of mediation techniques during the project. Often, mediation is one of the first alternatives to litigation and is an effort by the stakeholders to achieve an equitable settlement with the help of a skilled professional.

.8 Joint Recognition of Changes

One of the best ways of reducing potential claims is for the other party to recognize when a change has occurred. The common tendency *not* to do this, or to argue over every potential change, is a major factor in perpetuating claims.

Both parties need to be realistic. A clear change order process has to be implemented and incorporated in the contract to manage the changes, both to the scope of work and changes outside the scope.

.9 **Documentation.** Good documentation can lead quickly to acknowledging a change, whereas poor documentation will only prolong any argument between the parties. Good documentation can also provide a good defense against claims. A complete factual analysis can defeat spurious or poorly supported claims. See also Section 10.3.2 on Performance Reporting: Tools and Techniques of this extension, for effective processes that can lead to good documentation essential to timely and successful claim resolution.

16.3.3 Claim Prevention; Outputs

.1 **Changes**
Potential claims for compensation, or requests for extensions of time, or both, that are agreed upon, are turned into changes, thus eliminating the claims. Using the tools and techniques in Section 16.3.2 of this extension, there are no disputed requests for changes (claims) at the end of the project as they have been either disposed of as changes or withdrawn.

.2 **Stakeholder Respect and Acknowledgment**
Stakeholders who have successfully prevented or resolved potential claims, without the added time and expense of experts and attorneys, generally develop an appreciation and respect for each other. This, in turn, can lead to subsequent collaborative working relationships and future project successes.

16.4 Claim Resolution

Even with a concerted effort to prevent claims, they still may occur. There may be an understandable disagreement as to whether or not the claim in question is a change to the contract, or whether the claimed amount of compensation or time requested is correct or not. When this situation arises, a step-by-step process is set in to motion to resolve these questions. It is axiomatic that the longer this process takes, the more expensive and disruptive it is to both parties. Therefore, the goal is to settle these issues soon and at the lowest point in the organization as practicable. The process begins with negotiation, perhaps at more than one level, before moving on to mediation, arbitration, or litigation, depending upon the remedies afforded by the contract. Alternative methods of resolution have been increasingly used because of the proliferation of claims in construction and the expense of litigation. These alternative methods, called ADR for alternate dispute resolution, may include mediation, arbitration, and mini-trials.

16.4.1 Claim Resolution: Inputs

.1 **Statement of Claim**
See Section 16.1 of this extension.

.2 **Claim Quantification**
See Section 16.2 of this extension.

.3 **Contract**
The contract provides the ultimate baseline and means for resolution.

.4 **Fully Documented Certified Claim**
See Section 16.2.3.4 of this extension.

.5 **Expert Knowledge**
Construction personnel are often served well by seeking advice from experts in the area of construction claims. These professionals can often quickly assess the potential claim along with the existing documentation and provide knowledgeable information regarding the validity of a potential claim, existing documentation to substantiate it, and an order of magnitude in terms of cost and effort to fully document and pursue the claim. These professionals often can work directly with one party's attorney to combine both the technical and legal advice.

16.4.2 Claim Resolution: Tools and Techniques

.1 **Negotiation**
This is always the first and best step to resolution. Sometimes the negotiation needs to be elevated to a higher level but it still is a negotiation between parties trying to find an equitable solution.

.2 **Alternative Disputes Resolution (ADR)**
These include mediation, arbitration and mini-trials. Mediation is one of the first alternatives to litigation and is an effort by the stakeholders to achieve an equitable settlement by means of the use of a skilled professional mediator. Arbitration is just one step short of litigation and the judicial court system. This technique utilizes a trained professional, often an attorney, ex-judge, or construction expert, to act as the judge and jury to listen to stakeholder arguments, testimony, and assess factual and expert exhibits, to pass judgment, and make an award either in favor of the claim or in opposition to it.

.3 **Litigation**
This is the usual result when all earlier attempts at settlement have failed. Construction lawsuits are commonly complex for a jury to understand and often take a long time to present. This is considered to be a last resort and is expensive in terms of cost and upset to the organizations involved. Parties entering into litigation should be sure that this is the only way the dispute can be resolved.

.4 **Estimated Cost of Resolution**
When the initial attempts at negotiation fail, it is prudent for each of the parties to estimate the cost of carrying the dispute further. Mediators are costly (but can be cost effective) and some arbitration cases can approach the expense of litigation, due to the amount of discovery involved. An estimate of these costs can help in deciding just how important it is to keep pursuing a claim. Sometimes

the cost of pursuing or defending the claim can be considered a tradeoff and used as part of the settlement funds to resolve the claim prior to litigation.

.5 Claim Assessments or Expert Reports

Professional construction consultants can be hired to assess the issue and, in some cases, provide an expert report outlining the cost, time, and business impact of the potential claim. The consequential time and cost impact on other activities as a result of the extra or delayed work is important information that must be evaluated in terms of potential recovery, as some assessments will reveal documentation discrepancies and problems self-inflicted by the stakeholder which may prevent a full recovery. In some cases, an expert report may be obtained to corroborate the technical aspects of the claim.

16.4.3 Claim Resolution: Outputs

.1 Claim Resolved

This is one of the techniques covered in Section 16.4.2 of this extension, which settles the claim.

.2 Contract Closed

In cases where the contract cannot be closed because of a pending dispute, the resolution of that dispute enables the contract to be closed.

.3 Equitable Contract Adjustment

Usually in the form of a change order, an adjustment to the final contract amount is made as a result of the monetary settlement. The process of allocating this amount is usually an internal accounting function and varies depending on the type of public or private stakeholders involved.

Section IV

Appendices

Appendix A

Changes from Previous Edition of the Construction Extension

The purpose of this appendix is to provide a summary description of the most important changes to the *Construction Extension to a Guide to the Project Management Body of Knowledge* (*PMBOK® Guide*—2000 Edition) in order to create the current edition.

Alignment with *PMBOK® Guide*—Third Edition

The current edition of the *Construction Extension* was aligned in structure, style, and content with the *PMBOK® Guide*—Third Edition. As the two documents should be used together for construction projects, the alignment enables easier reference to the corresponding sections in each document. The process names and designations were updated to match the changes introduced in the *PMBOK® Guide*—Third Edition to enable consistency and clarity.

Emphasis on Emerging Industry Trends

The emerging issues identified during the review and update of the *Construction Extension* included emphasis on quality assurance and quality control, environmental issues, and project delivery methods. A solid emphasis towards the global aspects of the construction industry have introduced new considerations for the *Construction Extension*. International standards are now more prevalent and the reach of their application across borders creates new challenges and considerations for the construction industry. A recognizable issue that was not specifically addressed through this extension is the "green" construction trend that is influencing the way projects are conceptualized, designed, and constructed. This emerging trend has the potential for adding many new elements to future updates of the *Construction Extension* and is recognized here but not specifically addressed. The last emerging trend that was recognized is the human element and the demand for construction industry resources. As the industry market expands, the availability and skill base of the resources required to manage and execute construction projects will influence the way projects are organized, planned, and constructed. This will also create new elements for the *Construction Extension* in future updates that better address this human resource component.

Supplemental Construction Extension Chapters

The *Construction Extension* maintains the structure in Section III regarding the supplemental chapters:

Chapter 13—Project Safety Management
Chapter 14—Project Environmental Management
Chapter 15—Project Financial Management
Chapter 16—Claims Management

The team did not find it necessary to add any additional chapters but you can find much greater depth in the development of these areas and their consideration in the planning and execution of projects.

Conclusion

This update to the *Construction Extension* addresses, through the input and review by over one hundred industry professionals and subject matter experts, the elements and best management practices necessary to manage most construction projects with common applications. The new additions to the extension are highlighted in the process tables within each chapter. The extension should be used in conjunction with the *PMBOK® Guide*—Third Edition and viewed as a combined work. Due to the alignment with the *PMBOK® Guide*—Third Edition, the extension changes are best not viewed in comparison to previous extension editions but as a revised whole body of work that compliments the *PMBOK® Guide*—Third Edition.

Appendix B

Recognition of Specific Contributors

Specific recognition is given to Core Team Member Mr. Dermot Flood for his extraordinary contribution to the environmental and quality aspects of the extension and the extension update. From Ireland, Dermot was able to ignite an awareness in the team as to the importance and significance of the international standard practice and relevant global considerations that would have otherwise gone unaddressed. He truly was a catalyst in expanding the global view of the extension team and spent many hours preparing documentation and enlightening the team on how these issues impacted every chapter of the extension. We extend a special thanks to Mr. Flood for his efforts.

Mr. John Tocco, Esq. also deserves special recognition. His early participation in the areas of procurement, contracts, and administration were instrumental in the significant enhancements of these knowledge areas. Mr. Tocco's communication style and colorful graphics lead many discussions which needed a fair and balanced approach to resolution. He was able to guide us from strong opinion to responsible consideration of multiple viewpoints and the extension received great benefit from that. The team as a whole appreciates his input and contribution to the process.

Lastly, during the review of the exposure draft, the contributions by Hussain Ali Al-Ansari, Mohammed Safi Batley, and Mohammed Abdulla Al-Kuwari through the thoroughness of their review and insightful comments have given an added international dimension to the extension. Their efforts in reviewing and analyzing each chapter in depth gave the team a needed and valuable validation that the document was consistent, well thought through, and aligned with the *PMBOK® Guide*—Third Edition. The team would like to recognize and thank them for their extraordinary effort.

Jim Walker–*Construction Extension* Update Project Manager

Appendix C

Contributors to and Reviewers of the *Construction Extension to the PMBOK® Guide Third Edition*

The Project Management Institute gratefully acknowledges all of these individuals for their support and contributions to the project management profession.

C.1 The *Construction Extension to the PMBOK® Guide Third Edition* Project Core Team

The following individuals served as members, were contributors of text or concepts, and served as leaders within the Project Core Team (PCT):

James Walker, PMP, Project Manager
Syed Aqeel Kakakhel, PMP, Deputy Project Manager
James N. Brooke, PhD, PMP
Jeffrey S. Busch, PMP
Dermot Flood
Kelly C. Griffith, PMP
Paula M. Dolliver-Marshall, PMP
Kelly A. McBride
William A. Moylan, PhD, PMP

C.2 Significant Contributors

In addition to the members of the Project Core Team, the following individuals provided significant input or concepts:

Hussain Ali Al-Ansari
Mohammed Abdulla Al-Kuwari
Mohammed Safi Batley
Jean-Stephane Bedard, ING, PMP
Henry Hattenrath
Michael J. Jirinec, MSM, PMP
William G. Simpson, PE, PMP
John V. Tocco, Esq.

C.3 The *Construction Extension to the PMBOK® Guide Third Edition* Project Team Members

In addition to those listed previously, the following Team Members provided input to and recommendations on drafts of the *Construction Extension to the PMBOK® Guide Third Edition*:

Yasser Afaneh, PMP, M.Sc.
Stephen Kwasi Agyei, PMP, MRICS
Michael Atassi
Lee Barco, PE, PMP
Abhi Basu
Carlos Eduardo M. F. Braga, PMP
Michael J. Bennick
Mamoun Besaiso, C.E
Bernardo Bustamante
Juan Ignacio Boza Castro, MEng, PMP
Ricardo B. Cisco, PMP
Thomas Coalson
Rimple Dave, PMP
Edward E. (Ted) Douglas, CCC, PSP
Abdul Razzaq Edouni, DPA, PMP
Ahmed El Antary, PPM, MSPM
Patricia Ewanski
Erich E. Fox, PMP
Cristiane Fraga da Silva, PMP
Brad Garwood, MBA, PMP
Jamia Donett Grayson
Joseph A. Griffin, PMP, MBA
Luis Ruben Quintana Guevara
Jerome Guyard, MBA, PMP
Akkiraju V Harshavardhan, PMP
Ivan Hinkson PMP, CCC
Elly Hoinowski, PMP, MBA
Ir. Charles Victor Hunt
George A. Huser, Jr.
Mamane Ibrahim, PMP, APMC
Tony Jacob, PMP
Rohit Jain
Rick D. Johnson, AAIA, RB
Bijay Joseph
Sivamani Kanagasabai
S. N. Krishna Murthy
Ming Lee
Nicholas A. Letourneau, MBA, PMP
Simon Lewis
Antonio Quiroz Magan, MEng, PMP
Vladimir G. Maslev
Muhammad Aslam Mirza, MBA, PMP
Donald M. Missey, MBA, PMP
Brian G. Mitchell, MAATO, PMP
Sunu Mohan PMP, PSP
Maha M. Nassar, MBA, PMP
Naresh Chandra Nirmal, PMP
Valerie O'Keeffe-Short, PMP, CHAM
Michael J. O'Sullivan
Nasir Poonawala
Javier G. Puente T., PMP
Sameer Raleraskar, PMP
Gustavo de Abreu Ribas, MBA
Itza Maria Rivera
Michael M. Soliman Rofael
Fabrizio Rossi, MEng.
Raheel Sajid, MBA, MS
Shady R. Sammy, PMP, CDT
Satheesh Santhanagopalan
Hanumantha Rao Saripalli, DFSS-IT
Victor A. Singh, PMP
Brian Smith, Peng, PMP
Timothy Stephenson, PMP, OHST
Tejas V. Sura, MS, PMP
Aaron J. Sweeney, PMP, MPM
Ignacio Tanaka Murakami, PMP
Ian Trotman
Bhuvaneswaran V.
Cherian Varghese
Dharmendra H. Varma
Yuhong Wang
Zhiming Wu
James G. Zack, Jr., PMP, CCM
Salman Zakaria, PhD

C.4 Final Exposure Draft Reviewers and Contributors

In addition to team members, the following individuals provided recommendations for improving the *Construction Extension to the PMBOK® Guide Third Edition:*

Hussain Ali Al-Ansari
Mohammed Abdulla Al-Kuwari
Mohammed Safi Batley
Mamoun A. Besaiso, Esq
John E. Cormier
Theodore Creed, PE
Yazan A Darwazeh, PMP
Jim Delrie, PE
Romeo Diswe
Kashinath Raghunath Dixit
Charles T. Follin, PMP
Marian J. Hall, PMP
Robert B. Hierholtz ,PMP
William V. Hubbard, PE, PMP
James Ivey
Dennis Arthur Krenz, PMP
Stephen B. Lafferty
Leandro G. Martins
Jerry L. Partridge, PMP
Kenyon D. Potter, PE
Mario Salmona
Patricia C. Sibinelli, PMP
Joseph Henry Sites, PE
Robert Springmann
Glenn W. Strausser, PMP
Pieter van der Knaap, Sr.
Rebecca Ann Winston, Esq

C.5 PMI Project Management Standards Program Member Advisory Group

The following individuals served as members of the PMI Standards Program Member Advisory Group during development of the *Construction Extension to the PMBOK® Guide Third Edition*:

Julia M. Bednar, PMP
Douglas Clark
Terry Cooke-Davies, PhD, BA
Carol Holliday, PMP
Thomas Kurihara
Debbie O'Bray, CIM
Asbjorn Rolstadas, PhD
David W. Ross, PMP, PgMP
Cynthia Stackpole, PMP
Bobbye Underwood, PMP
Dave Violette, MPM, PMP

C.6 Production Staff

Special mention is due to the following employees of PMI:

Dan Goldfischer, Editor-in-Chief
Ruth Anne Guerrero, PMP, Former Standards Manager
John Zlockie, Standards Manager
Dottie Nichols, PMP, Former Standards Manager
M. Elaine Lazar, Standards Project Specialist
Roberta Storer, Product Editor
Kristin L. Vitello, Standards Project Specialist
Paul Wesman, Editor
Barbara Walsh, CAPM, Publications Planner
Nan Wolfslayer, Standards Project Specialist

Section V

Glossary and Index

Glossary

1. Common Acronyms

ADR	Alternate Dispute Resolution
DBOM	Design-Build-Operate-Maintain
DBOO	Design-Build-Own-Operate
DBOT	Design-Build-Operate-Transfer
EPC	Engineering-Procurement-Construction
EPCC	Engineering-Procurement-Construction-Commissioning
EPCM	Engineering-Procurement-Construction Management
RFI	Request for Information

2. Definitions

Activity Weights. A value assigned to activities, often in terms of worker hours.

Alternative Dispute Resolution (ADR). Methods, other than litigation, for resolving disputes including arbitration, mediation, and mini-trials.

Consortium. A group of companies formed to undertake a joint project.

Contractor. An individual or a company (commonly referred to as the seller), who is responsible for providing all of the resources necessary to manage and perform the work in the contract documents. The contractor may choose to subcontract work to other entities such as contractors with specialized expertise, material and equipment vendors, and testing services.

Contract Documents. Documents that consist of an agreement between owner (client) and contractor, and include conditions of the contract, drawings, specifications, and other documents listed in the agreement.

Constructability. The ease, safety, economy, and clarity of construction of a project.

Constructability Review. A review performed by personnel with expert knowledge of projects (or project components) that are similar in size, cost, and complexity, for purposes of assessing or determining: (a) whether the work can be performed with available means, methods, and resources while complying with the established schedule phasing, quality requirements, or (b) whether specialists are required, or (c) whether an alternative design is required. Constructability reviews usually incorporate value engineering.

Currency Hedging. A way of limiting exposure to future changes in the exchange rate of currencies.

Delivery Systems. Various methods of performing design/construction projects such as design-bid-build and design-build.

Design-Bid-Build. Design is completed by a professional architect or engineer; a construction contract is awarded after competitive bids.

Design-Build. A contracting method where the contractor is responsible for all aspects of the design and construction of the product in the contract documents. The scope of the contract includes management and design services; preparation and execution of construction documents; and construction, testing, and commissioning of the product.

Design-Build-Operate-Maintain (DBOM). Similar to DBOO except that the design builder has no ownership of the project.

Design-Build-Operate-Transfer (DBOT). Similar to DBOO except that the design builder will operate the facility for a period of time and then transfer ownership to another entity, for example highway tolls which are transferred to the state.

Dispute Review Board. A board formed at the start of or early in the project to review and adjudicate any disputes that may arise.

Eichleay Formula. A method used by the U.S. government for calculating overhead related to certain changes.

Exit Interviews. Interviews of construction (and project) staff as they leave the project as a means to record lessons learned.

Feasibility Study. An early engineering and financial analysis of a proposed project to determine its viability.

Force Majeure. An event not reasonably anticipated and acts of God such as weather, strikes, or other uncontrollable events.

Fringe Benefits. Costs of labor beyond wages. Such items may include as vacation, holidays, insurance, and taxes.

General Contractor. A contractor who does not specialize in one kind of work. Often used to refer to the major contractor who employs specialty subcontractors.

Hazard Analysis. A review of all the safety hazards that may be encountered in a project. Used to develop the safety and environmental plans. Also used to undertake safety and environmental risk analyses.

Inductions. Similar to tool box meetings, but used to convey specific job site practices or other pertinent issues to the field supervisors and workers.

Job Descriptions. A description of the responsibilities and authorities of an employee.

Joint Venture. A partnership of two or more engineering, construction, manufacturing trading, or investing companies often of limited duration.

Layout Risk. The risk associated with the design of the physical layout of a project.

Liquidated Damages. A requirement in contract documents for the buyer's recovery of estimated expenses from the seller that result from the seller's delay in meeting contract performance milestones. The estimated expenses typically include the buyer's anticipated cost for using alternate facilities or maintaining existing facilities, rescheduling or paying idle workforce, or for lost revenue.

Lump Sum Contract. A contract that is based on a fixed price amount for the work in the contract documents (see **Firm-Fixed-Price Contract**).

Non-Conformance Report. A report detailing the failure to meet specifications and often recommending a method of correction.

Non-Recourse. A type of finance that relies on the project only as lending collateral.

Partnering. A process, outside of the jurisdiction of contract documents and project plans, implemented by the project manager to motivate project participants assigned to a project. The purpose of the process is to obtain a buy-in and commitment from the project participants to ensure success. The process focuses on the unique benefits that the project's success will have on the individuals and the companies they represent.

Partnering (Alliance). Alliance partnering is a long-term relationship between an owner and an engineer/contractor whereas the contractor acts as a part the owner's organization for certain functions.

Partnering (Project-Specific). An informal agreement of all major entities in a project to work closely and harmoniously together.

Pre-Estimating Survey. A survey of a construction site to determine relevant characteristics such as weather, local suppliers and contractors, and available utilities.

Pre-Qualification List. A list of contractors or designers that have been preselected for further consideration based on their submitted qualifications.

Prime Contractor. A contractor holding a contract directly with the owner.

Progress Curves. A plot of a project's progress shown in percent complete versus amount of time. Used to display status and trends.

Progress Payments. A method defined in contract documents that specifies the payments to be made which correspond directly to the seller's monthly progress of work.

Project Specifications. The engineering and architectural plans and written requirements for a project. Similar to statement of work.

Punch List. The work items that are identified during a final inspection which need to be completed.

Recourse. Financing that is based on the assets of the sponsoring entity for collateral.

RFI. Request for Information. Typically a communication used by a contractor to request information or clarification from the designer or owner.

Self-Performed. Construction work that is performed by the major contractor's workforce.

Sensitivity Analysis. Varying several constituents of a calculated study to see what the effect is. Usually performed in connection with a feasibility study.

Short List. A list that is distilled from a larger group of proposers or bidders through the use of a set of criteria.

Sole Source. A type of procurement where only one supplier is asked to bid. Often required to obtain proprietary products.

Subcontractor. A contractor who is holding contract with a prime contractor (also referred to as a first tier subcontractor) or is holding a contract with a subcontractor to the prime contractor (i.e., lower tier subcontractor).

Substantial Completion. A contract milestone that is achieved by the owner's acceptance of the product constructed by the prime contractor. This milestone results in the owner utilizing the product for its intended function and purpose, and in generating a list of remaining items to be reworked, or of incidental items that do not affect the owner's use of the product to be completed. In some contract documents, this milestone terminates the accrual of liquidated damages for delays by the contractor in meeting performance milestones in the contract documents.

Tool Box Meetings. A regular meeting of field supervisors and workers to review important work issues; particularly those pertaining to safety. Tool box meetings are usually restricted to a specific subject, for example, excavation, concrete placing, or heavy lifts, etc.

Trades. Workers in the various construction disciplines such as carpenters and ironworkers.

Turn Key. A type of design build project where the design builder does all functions including start up before turning the project over to the owner.

Unit Rate Contract. A contract for construction based on established (bid) prices for certain types of work where the final quantities may not be known.

Value Management. Value engineering.

War room. A room used for project conferences and planning, often displaying maps, charts of cost and schedule status and other key project data.

Index